Future Perfect

Future Perfect

Confronting Decisions About Genetics

Lori B. Andrews

COLUMBIA UNIVERSITY PRESS   NEW YORK

COLUMBIA UNIVERSITY PRESS
Publishers Since 1893
New York Chichester, West Sussex
Copyright © 2001 Lori B. Andrews
All rights reserved

Library of Congress Cataloging-in-Publication Data

Andrews, Lori B., 1952–
    Future perfect: confronting decisions about genetics / Lori B. Andrews
        p. cm.
    Includes bibliographical references and index.
    ISBN 0–231–12162–8
    1. Genetic screening—social aspects. I. Title.

    RB155.65 .A53 2001
    616'.042—dc21   00–059657

Casebound editions of Columbia University Press books
are printed on permanent and durable acid-free paper.
Printed in the United States of America

c 10 9 8 7 6 5 4 3 2 1

For Christopher,
who is much more than the sum of his genes

Contents

Acknowledgments

I owe a great debt of gratitude to Nanette Elster, Dheanna Fikaris, Valerie Gutmann, Michelle Hibbert, Ami Jaeger, Gregory Kelson, Andrea Laiacona, Katherine Mason, Laura Swibel, and Lisa Wegryzyn for their advice and analysis, which created the framework for this book.

Like decent science, good policy cannot be done without building on the work of many previous scholars. Hats off to the especially talented George Annas, Arthur Caplan, Ellen Wright Clayton, Robert Cook-Deegan, Eric Juengst, Marc Lappé, Max Mehlman, Mark Rothstein, and Nancy Wexler, all of whom have challenged and aided me.

And a special thanks to the National Institutes of Health's Center for Human Genome Research (now the National Human Genome Research Institute) for grant R01–HG01277–01, which allowed me to pursue this project, and Chicago-Kent College of Law, which provided support and remarkable colleagues while I undertook it. And to Kate Wittenberg of Columbia University Press, who nurtured this book from its inception.

Future Perfect

1   Genetics Enters Our Lives

On a crisp fall day in 1995, Francis Collins, head of the Human Genome Project at the National Institutes of Health, took the microphone at a press conference he had convened. He announced that he and colleagues at Hebrew University in Jerusalem and at the University of California, San Diego, had discovered a mutation in a breast cancer gene in Ashkenazi Jewish women—the 185delAG mutation—that put them at higher risk than other women for developing breast and ovarian cancer.[1]

The efforts to discover genetic causes for breast cancer preceding that announcement could be presented as a scientific adventure story, with all the intrigue and excitement that accompanies any new biomedical development. Long years of research, scientists competing to be the first to claim a discovery, and squabbles over credit have been part of the story of the enterprise of genetic science since long before James Watson, codiscoverer of the structure of DNA, first described them in *The Double Helix*.[2]

The breast cancer saga was no different. Since 1990, when Mary-Claire King first pinpointed the general location of a gene that, when mutated, appeared to predispose women to the disease,[3] researchers had battled to find the elusive gene. When the first breast cancer gene was finally located, two biotechnology companies and the National Institutes of Health battled for scientific credit—and commercial patent rights that would ensure royalties each time a diagnostic test or treatment was undertaken with that gene. Once those battles were over, other researchers vied to be the first to find particular mutations in the gene that caused a higher-than-average risk of cancer.

Collins's announcement may have signaled the end of a scientific adventure story, but it was the beginning of a policy one, for each time a scientific quest ends in the genetics field, a policy quest begins. Health care providers, medical and scientific organizations, social institutions, legislatures, courts, and ordinary people seeking health care all struggle with the question of whether use of the new genetic technology should be encouraged or discouraged and what the social impact of that decision will be.

The evolving technologies for genetic diagnosis, treatment, and research raise profound questions. What is the psychological impact on the person who learns through genetic testing that he or she is likely to develop a serious disease later in life? Which, if any, social institutions—public health officials, insurers, employers, schools, or families—should have access to such genetic information? Should genetic interventions be used to enhance characteristics such as height or intelligence in "normal" individuals and to track antisocial behavior in individuals with "errant" genes? Should gene therapy be undertaken on embryos, changing their genetic inheritance, as well as that of future generations? Should cloning research be done to learn how genes are turned on and off—and should entire individuals be cloned for the ultimate genetic study of the impact of nature versus nurture?

The answers to those questions will determine not only the types of genes our future children will inherit but also the culture and values we will pass on to them. Genetic technologies could be applied in attempts to create "perfect" people. But moving in that direction entails such high psychological and social costs that the society we would be creating would be far from perfect.

The Rapid Integration of Genetic Tests Into Clinical Practice

Genetic discoveries are entering our lives at an astonishing pace. The day after Francis Collins's press conference, newspapers around the country reported the discovery of the 185delAG mutation as front page news. In response, numerous women called their doctors asking whether they should get tested. Biotechnology companies geared up to provide the test. At the time of Collins's press conference, he and the other researchers had identified only *eight* women in the study with the genetic mutation.[4] All that was known was that these women had a glitch in their breast cancer gene. What the glitch meant was still a mystery. It was unknown, for example, whether women with this glitch would actually get breast cancer—or whether the genetic aberration was merely a variation of the normal gene. Much additional research was needed. But since the test for this unusual pattern in the gene is very easy for any lab to perform, it was offered to women before important fundamental questions about its meaning, usefulness, and desirability were answered.

The women who sought 185delAG testing shortly after this announcement were told that if they had the mutation, they had an 87 percent chance of getting breast cancer. Some women whose test results indicated that they had the mutation decided to undergo "prophylactic" mastectomies and have their healthy breast removed to try to avoid the cancer. Later research found that far fewer women with the mutation—50 percent—would develop cancer than initially thought. Some women may have had their breasts removed unnecessarily.

Bernadine Healy, the director of the National Institutes of Health when Francis Collins made his announcement, later criticized the rapid movement of genetic tests for breast cancer into clinical use. "The race to discover new genes was visible and competitive, the promise of substantial economic returns for screening tests compelling, and the public hunger for a breakthrough in breast-cancer treatment intense," noted Healy in 1997, after the new data were revealed suggesting that fewer women with the mutation would get cancer than previously thought. "It is too early to use BRCA [breast cancer] gene testing in everyday clinical practice, because it violates a common-sense rule of medicine: don't order a test if you lack the facts to know how to interpret the results."[5] It is still unclear which breast cancer gene mutations will actually lead to breast cancer—and it is unclear whether women will actually benefit from genetic foreknowledge. Women who undergo prophylactic surgeries still run the risk of developing cancer in their remaining tissue. Additionally, being identified as having a genetic mutation may cause them to be stigmatized and subjected to discrimination.

The saga of the breast cancer gene is repeated daily as new genes and their mutations are discovered. The much-touted Human Genome Project has led to an increase in the number of genetic services available. Initiated in 1990 by the U.S. Congress, this fifteen-year project is attempting to map (determine the location) and sequence (determine the chemical makeup) of all the 80,000–140,000 genes in the human body. This and related research activities have led to a torrent of reports in the scientific literature and the mass media about genes that predict an increased likelihood that an individual will have a particular disorder or characteristic. Barely a week goes by without a report of the sighting of a new gene, leaving both the public and the policymakers in a quandary over what to do with the new information.

As with breast cancer testing, a similar quick transition from research to practice occurred when initial, limited research indicated that having one or two APOE 4 alleles of the apolipoprotein gene might indicate a genetic propensity to Alzheimer's disease. The test is far from perfect. Many people with the APOE 4 allele will never develop Alzheimer's disease, and many people with the disease do not have that allele.[6] Thus the test cannot predict whether an individual will get Alzheimer's.[7] It can be used only as a diagnostic tool for those who already show signs of dementia. As with the breast cancer gene, numerous questions still need to be answered, including questions about how the test works in different racial groups, particularly since the genetic mutation that seems to increase the chance of Alzheimer's in whites does not have the same effect in African Americans.[8] Though a wide variety of professional organizations have issued statements recommending that APOE testing not be offered to people outside of a research setting, when one East Coast researcher tried to recruit people to undergo genetic testing for a study, he had difficulty finding elderly people in New York City who had not been tested. In contravention of the recommendations, doctors had already begun testing their patients.

Increasingly, people are tested for diseases that will not manifest until later in life, a practice that is creating a new class of individuals, referred to as the asymptomatic ill, who, to all appearances, seem healthy. For some diseases, having a genetic mutation means that the person will almost certainly develop the disease. But for others, genetic mutations only slightly increase the probability of falling ill. Such uncertainty makes it difficult for the person to determine how to incorporate this new information into his or her life.

Prenatal genetic testing—to assess the health status of a fetus—has been around for nearly three decades, but even that is changing. Now such testing is possible not only for serious disorders[9] but also for less serious disorders, diseases that are treatable after birth, for disorders that do not manifest until later in life, such as Huntington's disease and breast cancer, and even for conditions that are not medical problems, such as homosexuality. Already couples have sought genetic testing for Alzheimer's for their fetuses, intending to abort even though the child could have seventy or eighty years of a normal life before manifesting symptoms.[10]

The meaning of genetic tests varies widely. There are single-gene disorders for which environmental factors or other genes may increase or decrease the likelihood that the disease will actually express itself. In scientific terms, such mutations are said not to be fully penetrant—that is, not everyone with the genetic mutation will develop the disorder. Often the gene mutation indicates only a predisposition to a problem, and it takes an additional intervention, such as a particular environmental exposure, to trigger the condition.

Even for disorders that are completely penetrant, it is impossible to predict how severe the disease might be or when it will strike. Even though the average age of onset for Huntington's disease is the forties, children as young as two have been symptomatic of the disease, while other people have not manifested symptoms until their late seventies.[11] Similarly, some people with genetic mutations (such as certain people with the cystic fibrosis mutation) have such a mild manifestation of the disease (or even no symptoms whatsoever) that it seems odd to think of them as having a disease. Others with the same genetic profile have serious health problems.

How Genetics Differs from Other Medical Realms

Genetics shares many features with other medical fields, but several of its characteristics heighten concerns about its application. First, genetics often affects central aspects of our lives. Because genes are usually viewed as immutable and central to the determination of who a person is, information about genetic mutations may cause a person to change his or her self-perception and may also alter the way others treat that person.[12] Second, there is a chance that people will get genetic testing or therapy without sufficient advance consideration of its potential effects. In most instances people seek medical services because they are already ill. But with predictive genetic testing, there may be an incentive for biomedical companies and physicians to market such tests heavily, and healthy people interested in learning whether they are at risk for later diseases may not consider the psychological, social, and financial impact of learning genetic information about themselves before they agree to genetic testing. As one group of cancer researchers observes with respect to genetic testing, "Society's technological capabilities have outpaced its understanding of the psychological consequences."[13]

There is also the problem of the therapeutic gap. Many diseases can be diagnosed through genetic testing, but few can be treated successfully. This has enormous social implications—your health insurer may drop you based on a genetic test result. But it also has potentially risky medical

RECESSIVE, DOMINANT, AND X-LINKED DISORDERS

Testing can be done for chromosomal disorders, such as Down syndrome. Tests are also available for single-gene disorders, caused by a defect in a particular gene pair. According to Dr. David Wheeler of the National Center for Biotechnology Information, there are 10,114 single-gene disorders. These disorders are of two types—dominant or recessive. The dominant single-gene disorders include achondroplasia (a type of dwarfism), some forms of chronic glaucoma (which causes blindness), Huntington's disease (which causes degeneration of the nervous system), and hypercholesterolemia (a high level of cholesterol in the blood that may lead to heart disease). The affected parent has one normal gene and one faulty gene in the specific pair causing the disease, but because the defective gene is dominant, the normal gene cannot compensate for the problems caused by the faulty one. Since the child inherits only one of the two genes in a pair from each parent, there is a 50 percent chance that the gene that parent passes on to the child will be a faulty one and the child, too, will suffer from the dominant genetic disorder. Sometimes, though, the problem will not manifest itself at birth but will appear much later in the child's life. The dominant single-gene disorder, Huntington's disease, for example, does not generally appear until the people affected are in their late forties.

In the case of a recessive single-gene disorder, the person who has one faulty gene in a pair is generally not affected by the disease because the normal gene in the pair compensates for the faulty gene. That person is known as a carrier. However, if two carriers have a child together, there is a one-in-four chance that the child will inherit one faulty gene from each parent and thus have the recessive single-gene disease. These include sickle cell anemia (a blood disorder that primarily affects African Americans), cystic fibrosis (a disorder affecting the lungs and pancreas), phenylketonuria, also known as PKU (a deficiency in an essential liver enzyme), galactosemia (an inability to metabolize milk sugar), thalassemia (a blood disorder that primarily affects people of Mediterranean ancestry), and Tay-Sachs disease (a disorder of the nervous system that affects mostly Eastern European Jews).

X-linked disorders, another type of genetic difficulty, affect males who inherit the defects from their mothers. In this situation, the defective gene occurs only on the X chromosome. These defects are generally recessive. Since every woman has two X chromosomes, there is a 50 percent chance she will pass on her defective X to her children and a 50 percent chance she will pass on her healthy X. The daughter of a woman with an X-linked disorder is generally unaffected. Because she gets an X from her mother and an X from her father, even if she gets the X with the mutation from her mother, its defects are made up for by the normal X her father gives her. Since a son always gets a Y from his father, an X with a mutation from his mother will *not* be compensated for and he will suffer from the disease. Over one hundred X-linked disorders are known. They include hemophilia (also known as bleeding disease), Duchenne muscular dystrophy (involving muscle deterioration), and a mental retardation in boys known as Fragile X.

implications. While genetic treatment and preventive strategies for asymptomatic individuals are being developed, positive results on a genetic test may lead to interventions that are costly, unnecessary, ineffective, or even harmful.

Genetics has another unique feature. Genetic testing of a particular individual reveals genetic risk information about his or her relatives as well. A parent and a child have half of their genes in common, as do siblings. Cousins share one-quarter of their genes, as do grandparents and grandchildren. Family bonds raise new and profound questions of "gen-etiquette," questions of the moral obligations that may emerge with the acquisition and disclosure of genetic information. If a woman learns she has a genetic mutation predisposing to breast cancer, does she have a moral or even a legal duty to share that information with an estranged cousin?

In Ventura County, California, a woman convinced the local hospital to allow her to do paternity testing on a blood sample from her recently deceased father. The test suggested that the man she'd called daddy all her life was not her biological father, indicating that perhaps her mother had an affair. Now her mother is suing the hospital for invasion of privacy—and seeking to have more-sophisticated tests done on the sample to show that the first results were wrong.

In the past, genetic testing has generally been used like other clinical testing—in situations in which the patient was symptomatic or the patient's family history, age, or ethnic background suggested a particular risk. Now genetic testing is being suggested for the population at large, to predict future diseases. The idea seems simple and seductive—look into your medical crystal ball, see your future diseases, and try to prevent them. The reality is much more complicated, however. For most complex disorders, genetic test results are ambiguous,[14] and prevention and treatment strategies are uncertain as well.[15] Some diseases, such as Huntington's, are currently incurable; for those who are afflicted, the future holds only certain debilitation and death. Learning the information may cause the person to give up on his future, or it may cause social institutions (professional schools, insurers, or employers) to give up on him.

Genetic technologies are entering our lives in a variety of ways, with profound effects. Yet there are few places that we can turn for objective advice about the advantages and disadvantages of using a particular genetic service.

Commercialism in Genetics

Every day thousands of people face individual choices about genetics. At the same time, hundreds of policy decisions are made about genetics—such as whether a particular test should be offered, what information should be provided in advance of a test, and who should have access to the results. Doctors determine whether to test patients for a newly discovered genetic mutation. Health insurers analyze whether to reimburse for such a test. The Food and Drug Administration struggles with the question of whether university laboratories offering genetic tests should be regulated. State lawmakers debate whether an insurer should be able to deny an individual health insurance because his or her sibling or parent has a genetic disorder.

Yet unlike any major medical dilemma we have faced in the past, we do not have a sufficient body of "neutral" scientists to advise us on these matters. A series of legal developments in the 1980s turned genetic science from a public interest activity into a commercial one. A landmark U.S. Supreme Court case in 1980 granted a patent on a life-form—bacteria—setting the stage for the patenting of human genes.[16] Initially researchers assumed that people's genes were not patentable, since patent law covers "inventions" and prohibits patenting the "products of nature."[17] But by the mid-1980s, the U.S. Patent and Trademark Office was granting an increasing number of patents for human genes, allowing the researcher who identifies a gene to earn royalties on any test or therapy created with that gene.[18] A second radical change in the 1980s was a series of federal laws allowing university researchers and government researchers to reap the profits from their taxpayer-supported research.[19] This development encouraged collaborations between researchers and biotechnology companies—and a growing interest in the economic value of genetic technologies.

In the context of advances in biotechnology, the 1980s legislation led to important changes in the goals and practices of science and medicine. Leon Rosenberg, when he was dean of the Yale University School of Medicine, described the influence of the biotechnology revolution on scientific research: "It has moved us, literally or figuratively, from the class room to the board room and from the New England Journal to the Wall Street Journal."[20] Thus, at the same time that genetic technologies are being increasingly marketed, fewer and fewer neutral geneticists are available to serve as advisers to society on the merits and impacts of these technologies.

According to a 1992 study by Tufts University professor Sheldon Krimsky, in 34 percent of 789 biomedical papers published by university scientists in Massachusetts, at least one of the authors stood to make money from the results they were reporting.[21] This was because they either held a patent or served as an officer or adviser of a biotech firm exploiting the research. None of the articles disclosed this financial interest, despite the fact that it could have biased the research.

Biotechnology companies and physicians heavily market genetic services and products, and the supposedly neutral scientists developing them often share a cut of the profits as patent holders or board members of such companies. The commercial incentives mean that marketing is aggressive and tests are applied clinically within months of the discovery of a particular genetic mutation. Genetic testing is rapidly being deployed as a tool in a variety of medical fields. Neurologists test patients for Huntington's disease; gerontologists for Alzheimer's; and internists for genetic propensities to early coronary disease. Even dentists are getting into the act, offering testing for a genetic propensity to periodontal disease, to determine which patients should receive antibiotics.[22] Genetics is touted as a way to revolutionize medicine. Its boosters predict an era of true preventive medicine, where genetic propensities for disease are identified and preventive strategies undertaken to forestall the manifestation of the disease.

The commercialized environment makes it more likely that tests will be implemented prematurely, that they will be performed without appropriate concern for informed consent, and that the poor and disadvantaged will be the least likely to share in any benefits.

Commercialization has pushed genetics into new settings. The genetics industry and the funeral industry have teamed up to market a kit that allows funeral directors to take some DNA from the deceased.[23] It is promoted by the manufacturer, GeneLinks, as a way to obtain genetic information about the deceased for the aid of other relatives, but in truth the need for such an approach has decreased as direct gene testing has replaced the need for family linkage studies.[24] Rather, the company is cashing in on the glamour of genetics, creating new psychological needs and then meeting them. Some families say they find it comforting to have a bit of their dead relative around in the form of his or her DNA in storage.

While GeneLinks sends its samples to the University of North Texas for storage and potential analysis, another company, GENE*R*ATION LINKS, allows you to keep DNA of yourself and family members in your basement.[25]

GENE*R*ATION LINKS markets a kit with which you can collect hair, tissue from a cheek swab, and (with a doctor's help) blood and store it. "Our philosophy is that your genetic information should be under *your control*. The home/self-storage Kit is designed to maintain privacy," says the company's brochure.

Even the Internet has become a site for genetic technologies. An Indiana doctor has a Website that allows people access to confidential, predispositional genetic testing for a variety of disorders, including breast cancer, Alzheimer's, and juvenile onset diabetes.[26] In addition, researchers who want to get blood samples for genetic research can buy them via the Internet from tissue repositories. In most instances, the patients from whom the samples came have no idea that the blood they provided for hospital blood tests and the tissue they provided for biopsies have been turned into profitable cell lines. In an extraordinary breach of privacy, one company even put identifiable patients' medical records on the Internet without their knowledge and consent, which allowed tissue buyers to learn more about the patients who were the sources of the cell lines.[27]

Simplifying Genetic Tests

Evolving, less-intrusive technologies for genetic testing make such tests simpler to do—which may give people the mistaken impression that the testing does not raise profound questions. A Q-tip can be used to swab the inside of the cheek to remove cells; this procedure may seem less threatening to some groups (such as children) than a blood test. A new prenatal technology, fetal cell sorting, provides information about the fetus without creating the physical risk to the fetus or the pregnant woman that is posed by amniocentesis or chorionic villi sampling (CVS). A blood test is performed on the woman, and complex procedures in the laboratory capture minute amounts of fetal blood cells that are circulating in the woman's blood.[28] Prenatal diagnosis on those cells can determine whether, for example, the fetus has Down syndrome,[29] cystic fibrosis (CF),[30] or Tay-Sachs disease.[31] Unlike amniocentesis and chorionic villi sampling, fetal cell sorting can be done surreptitiously. Blood is routinely drawn during pregnancy for a variety of legitimate purposes, and it would be simple to subject that blood to genetic testing without the woman's knowledge.

With these less-intrusive testing technologies, genetic testing is being done without advance consideration of the impact. Moreover, simpler technologies make it easier for other social institutions, such as courts, insurers, and schools, to require testing—or even to undertake it without the target's knowledge or consent.

Another technological development that is changing the nature of genetic testing is multiplex testing, in which numerous genetic tests can be performed on a single tissue sample.[32] Microchips are being developed that can test a person's blood sample for 200–300 genetic mutations at once, and that capability will soon be expanded to 5,000–10,000 mutations. This means that people will be offered genetic testing for a greater range of disorders, and it is less likely that they will receive information in advance of the testing, since health care providers will not have time to explain each disorder. Consequently, people will be tested for many diseases about which they have little knowledge; the way the health care providers describe the disorder may have undue influence on whether people choose to undergo testing and whether, if the test is a prenatal one, they choose to abort based on the results.

The dizzying assortment of available genetic services raises challenges for us as individuals and as members of a larger community. In the next few years, each of us will be faced with the question of whether to undergo genetic testing. In some instances, we may even find—as hundreds of people already have—that we have been tested without our knowledge or consent. Insurers, employers, or courts may be making decisions about us based on our genes.

In a case involving termination of parental rights, a South Carolina court ordered a woman to undergo genetic testing for Huntington's disease.[33] Insurers have denied coverage to currently healthy people because they have genes associated with a late-onset disorder.[34] In Colorado and Georgia, a genetic testing program for a form of mental retardation called Fragile X is used in the grade schools.[35] Since the genetic testing provides no additional information that could change how the children are handled, critics suggest that it will only serve to stigmatize the children.

Exaggerated Faith in Genetics

The drive toward testing is also spurred by an unquestioned belief in the predictive power of genetics. While scientists and doctors once thought microorganisms were responsible for all human ailments,[36] they now blame genes. This assumption influences not only the type of research that gets funded but also the way disease and abnormality get categorized. The result, says McGill University professor of epidemiology and biostatistics Abby Lippman, is "a process of colonization with genetic technologies and approaches applied to areas not necessarily—or even apparently—genetic."[37]

This "geneticization," as Lippman calls it, is pervasive, though not necessarily beneficial. Gene therapies are often sought by researchers even when conventional therapies might be more appropriate and more readily attainable.[38] Genetics is viewed as an appropriate way to guide everything from family dynamics to the legal system. While twenty years ago, child-rearing manuals focused on telling parents to enhance their child's environment, modern guides tell parents there's little they can do—"it's all in the genes."[39]

There are few places to turn for guidance about the use of genetic technologies. Those who favor the technologies have made excessive promises of the benefit to mankind, including the potential for dazzling medical achievements. Calls for regulation of genetics have met with resistance and claims that any limitations on research would mean that therapeutic advances would be lost or curtailed. Even universally recognized standards for research—such as informed consent requirements—are seen in some quarters as unduly hampering genetic progress.[40]

Researchers convince patients that their salvation lies just around the corner, in the next genetic development. Within two weeks after Ian Wilmut and Keith Campbell announced that they had cloned a mammal—before any research had been done to show any therapeutic benefits resulting from human cloning technologies—a cancer patient was telling reporters that President Clinton's moratorium on federal funding for human cloning research was preventing him from getting the one treatment that could save his life. A lawyer was arguing that if couples were not allowed to clone their dead child, their fundamental right to make reproductive decisions would be violated.[41]

On the other hand, critics who remember the "war on cancer"[42] and researchers' and politicians' promises that cancer would be cured by 1979 wonder why they should believe the current crop of zealous promises. They are even more skeptical of scientific claims now that medical research has become so commercialized, with government and university researchers discovering certain genes, patenting them, and then benefiting financially each time a gene test is performed. Other critics express concern that genetic testing and gene therapy will change fundamental social values—for example, leading to greater discrimination against the disabled. Yet the dire predictions of certain vocal critics—describing a genetic Armageddon[43] or a future of cloned despots[44]—sometimes obscure a close analysis of the actual benefits and risks of genetic technologies.

Charting the Future of Genetic Policy

A multitude of policy decisions are made each day about genetics. Some of these decisions are most appropriately made by health care providers or business organizations, while others may be better addressed by professional or trade organization guidelines or by formal laws or regulations. No matter at what level the issues are addressed, we all need better information about what genetic technologies can and cannot do.

In this book, I analyze the three frameworks by which most health care services in the United States are regulated: the medical model, the public health model, and the fundamental rights model. Then, in order that the reader may determine which framework is appropriate for genetic services, I describe the impact of genetic services on self-image, personal relationships, reproductive decisions, parenting, group identity, and treatment by social institutions, and discuss the problem of ensuring the quality of genetic services. This illustrates how the medical, public health, and fundamental rights approaches can influence the type of information we receive about genetic services, the control we have over whether genetic services will be used on us without our knowledge or consent, and the kinds of decisions that will be made about us based on our genes.

Genetic services are now emerging at a rate that challenges our ability to determine appropriate policy. Decision makers must catch up with these technologies to avert misuse of genetic information. In order to escape the myriad conflicting and overlapping policies that act only as Band-Aids to the problems, we must lay the foundation by choosing a framework that best suits the issues surrounding the use of genetic services. We must also avoid the simplicity of genetic determinism. In many cases, our genes only predispose us to certain traits or represent the probability that certain diseases will manifest. Even this information, however, can be dangerous without safety mechanisms in place that will allow neutral policymaking, educate the public, secure suitable quality assurance measures, and ensure voluntary informed consent.

In June 2000, researcher J. Craig Venter, president of Celera Genomics Corporation, along with Francis Collins, director of the National Human Genome Research Institute, announced the completion of a rough draft of the sequence of the entire human genome.[45] Ready or not, both as individuals and as members of society we are faced with momentous decisions about the uses and abuses of genetics.

2 Competing Frameworks for Genetics Policy

Genetic technologies raise issues that cut to the core of what it means to be human and what it means to be a just and fair society. Yet the significance of those technologies has not been addressed in a systematic way by policymakers in the United States. Instead, a chaotic series of pronouncements by various agencies, medical organizations, health care institutions, and legislatures has addressed narrow issues or isolated subcategories of genetic technologies. The result has been unnecessary duplication of effort, conflicting guidelines, and specialized policies that can cause harm when inappropriately applied more widely.

The United States notably lacks an adequate structural mechanism for analyzing controversial new medical technologies. In other countries, such mechanisms do exist. In some, licensing bodies determine when it is appropriate for clinics to begin to offer a particular service, such as genetic testing of embryos.[1] In others, a national commission is appointed to recommend policies to the legislature. In Canada, a royal commission was chartered to recommend policies governing genetic and reproductive technologies as a whole. The commission used a variety of innovative methods to address these issues. It instituted a toll-free telephone number so citizens could report their own experiences with these technologies and express their opinions. In order to assess the values that defined Canadian life, the commission sought research and analysis from representatives of seventy disciplines on such topics as the psychological and social impacts of infertility, assisted reproduction, human embryo research, genetic testing, and the use of fetal tissue. The commission determined that Canadian social values stressed non-commodification and non-objectification, as well as protection of the vulnerable. This led to the recommendation of bans on human cloning, paid surrogate mother-hood, genetic enhancement, and sex selection for nonmedical purposes.[2]

In the United States, however, the dominant social value can be described as "show me the money." As a range of genetic technologies, some of du-bious value, become available in various settings with little oversight, there are many reasons to be concerned. "It's very difficult to regulate, in part, because the core technology of dealing with genes is remarkably simple," says J. Craig Venter, who has sequenced the DNA from more than 40,000 genes. A few years ago, he let third graders sequence DNA at his institute. "The technology is now being taught in grade school. You don't need a Ph.D. for it. In fact it's probably easier to understand if you don't have one."[3]

Concerns about appropriate uses of certain genetic technologies have been addressed by a variety of professional and governmental entities, among them the Institute of Medicine of the National Academy of Sciences,[4] the National Institutes of Health/Department of Energy Working Group on the Ethical, Legal, and Social Implications Program of the Human Genome Project,[5] the National Bioethics Advisory Commission,[6] the (now defunct) Office of Technology Assessment (OTA) of the United States Congress,[7] and professional organizations such as the American Society of Human Genetics (ASHG)[8] and the American College of Medical Genetics (ACMG).[9] These entities have generally addressed each issue with ad hoc committees meeting for limited periods of time and with no mechanism for situating the subject they are addressing within a larger social context. In addition, these entities are usually reactive, springing into action once a particular technology has been developed. The discovery of the genetic mutation associated with cystic fibrosis, for example, led to a series of policy deliberations—by professional organizations, the National Institutes of Health, and the Office of Technol-ogy Assessment—about the use of the cystic fibrosis test.

These efforts have suffered from several deficiencies. Some groups have been subject to "capture" by the governmental or professional organizations that formed them and have not adequately represented the public. Other committees created guidelines in a vacuum without adequate staffing or any mechanism to collect data to assess the impact of their proposed guidelines. Still others addressed only the physical risks of proposed procedures, not the social values at stake. This overlooks the fact that studies find that "risk is less significant than moral acceptability in shaping public perceptions of biotechnology."[10] Consequently, the public does "not find the language of objective risk assessment adequate, arguing that risks are fundamentally moral and political."[11] Moreover, previous efforts often failed to address the real-life situations in which people make decisions about genetics. As an example, the Office of Technology Assessment report on genetic testing for cystic fibrosis provided elaborate economic models about the cost savings to society if women aborted fetuses affected with cystic fibrosis. However, the models ignored data about how women make such decisions and what proportion of women pregnant with fetuses affected with cystic fibrosis would abort.[12]

Policy deliberations have generally[13] focused on a particular technology (such as cystic fibrosis or breast cancer testing),[14] a particular subject of technology (such as children),[15] or a particular recipient of genetic information (such as insurers).[16] The resulting reports have not assessed how a decision on that particular subject could affect other uses of genetic technologies. The scholarly articles, too, usually focus on a particular application of the technology or a particular legal or ethical issue. Consequently, there has been little attempt to create an overall conceptual framework within which to regulate genetics.

Such a framework might be helpful for several reasons. First, it would provide an opportunity for a better understanding of the values that undergird professional and public policies in this area by allowing a closer scrutiny of the principles underlying specific policies. It could transform what may seem like an isolated individual case (Should Aunt Millie be told that she, too, might have a gene predisposing her to breast cancer?) into a larger category of cases (To what extent should we facilitate people's learning about their genetic makeup?). Additionally, by assessing how alternative conceptual frameworks would address genetics issues about which policies have not yet been adopted, a preexisting framework would provide a basis for developing forward-looking policies that could readily be invoked to handle new technologies or new issues, rather than relying on reactive policies to deal only with current crises.[17]

When technologies are introduced incrementally and policies are adopted in small units to deal with a few isolated issues, there is less opportunity to stimulate a social debate about whether we are moving in a direction in which we wish to go.[18] Benjamin Wilfond and Kathleen Nolan criticize what they label the "extemporaneous" manner in which genetics policy has been made in the United States.[19] They argue in favor of using an "evidentiary" approach incorporating an evaluation of research and attention to underlying normative issues. Their focus is on "clinical" research, but I think it is also appropriate to consider research from a wider variety of disciplines, such as anthropology, sociology, philosophy, history, and linguistics.

The need for a framework with respect to genetics policy issues was raised by Robert Blank, in his book *Regulating Reproduction*, where he noted that "by concentrating on one or several applications, the cumulative impact of reproductive and genetic technologies is obscured." Blank observes that a "fragmented policy-making process and its tendency to focus on immediate, conspicuous problems has led to a failure to provide systematic, comprehensive assessment of the technologies or their implications for society."[20]

Yet at the very time that public input and overall deliberations about genetic decisions are most needed, the government is cutting back on oversight.[21] The Food and Drug Administration regulates genetic tests sold as kits—which does not cover the vast majority of tests, which are sold as services by university, hospital, and biotechnology laboratories.[22] Even the most controversial technology—changing the blueprint of life itself through gene therapy—is given cursory regulation. With respect to gene therapy, National Institutes of Health director Harold Varmus cut the powers of the interdisciplinary Recombinant Advisory Committee significantly.[23] That group met in public, considered the ethical as well as the scientific merits of proposals, and insisted on adequate informed consent. Now the Food and Drug Administration will have the prime responsibility for reviewing gene therapy proposals[24] and can make those decisions in secret, without concern for their larger social impact. In contrast, in England, gene therapy protocols are reviewed by a statutory Medical Controls Agency, similar to the FDA, as well as the governmentally-created Gene Therapy Advisory Committee (GTAC).[25] The dangers of the more lax approach in the United States were underscored dramatically in late 1999 when a research subject died as a result of an injection of experimental gene therapy. In the wake of the death, it was learned that gene therapy researchers had dramatically understated the potential untoward side effects of the experimental therapy. Only 39 of the 691 deaths and illnesses suffered by gene therapy recipients given a similar vector to the one that caused the death were reported to the National Institutes of Health as required by the federal rules.[26]

A more systematic approach to addressing genetic technologies would be to develop a central set of principles to apply to all such technologies, to serve as a starting point. Any deviation from those principles would then have to be justified by sufficient evidence and analysis.

But how can such principles be derived? A promising approach is to analyze the three different conceptual frameworks that govern the adoption of other medical technologies in the United States (the medical model, the public health model, and the fundamental rights model) to determine which one is the most appropriate for genetics.

Historical Context of Genetics

The need to tread carefully when developing policies for genetic technologies is underscored by its historical context. Both law and science have taken a number of false turns in incorporating genetics into their respective disciplines. In the late 1800s, a majority of the geneticists in this country believed that one could extend genetic principles to explain human behavior.[27] Traits such as feeblemindedness,[28] criminality,[29] pauperism,[30] prostitution,[31] and seafaringness[32] were all thought to be single-gene defects. In fact, seafaringness—the desire to be a sailor—was thought to be an X-linked trait because it seemed to affect only men.[33]

During the early United States eugenics movement, genetic explanations were used for political purposes. They were based on prejudice[34] and used to try to maintain the economic positions and values of those in power in society.[35] Anyone who did not share those values—or who represented an economic threat—was categorized as having bad genes. This effort consequently had a disproportionate impact on the least powerful and most disadvantaged in society—women, minorities, and the poor. Women who had children out of wedlock were labeled feebleminded and were sterilized.[36] Men who protested working conditions were said to have lesser genes.[37]

Genetic theories quickly served as the basis of proposals for social and legal reform. The prime thrust of the reforms was to prevent people with presumably undesirable genes from reproducing. In the late 1800s, the chairman of the psychology department at Harvard advocated "the replacement of democracy by a caste system based upon biological capacity with legal restrictions upon breeding by the lower castes and upon intermarriage between the castes."[38] Federal and state legislatures took the teaching of genetics to heart. They passed laws to prevent people with presumably undesirable genes from reproducing, on the grounds that the care of the unfit (such as the mentally disabled) was draining society's resources.[39]

The public policy of sterilization was adopted with the belief that it would prevent future genetic disease and save the rest of society money. In the

1870s, state governments had provided extensive funding for institutions for the care of the feebleminded, but subsequently they began reassessing this expenditure.[40] As it pressed for mandatory sterilization laws, the American Eugenics Society pointed out that the descendants of one allegedly genetically inferior pauper couple—the Jukes—had cost the State of New York $2 million, but that it would have only cost $150 to sterilize the original Jukes pair.[41] At a time when attention was focused on gangsters, the American Eugenics Society told the public that crime, a function of hereditary defects, was costing the average family $500 annually.[42] In the 1920s, county fairs exhibited a display that "revealed with flashing lights that every fifteen seconds a hundred dollars of your money went for the care of persons with bad heredity, that every forty-eight seconds a mentally deficient person was born in the United States, and that only every seven and a half minutes did the United States enjoy the birth of 'a high grade person . . . who will have the ability to do creative work and be fit for leadership.'"[43]

The first eugenics law, enacted in Indiana in 1907, provided for the involuntary sterilization of institutionalized, unimprovable individuals who were idiots, imbeciles, rapists, or habitual criminals. By the end of the next decade, sixteen other states had followed suit.[44] More than 60,000 people were sterilized under those laws.[45] In Germany, the Nazis modeled their sterilization law after the American one.[46] And the German example was used by some American eugenicists in an attempt to spur on U.S. lawmakers. In 1934 a doctor tried to get the Virginia legislature to broaden its sterilization law by arguing: "The Germans are beating us at our own game."[47]

The Nazi misuse of American sterilization did not dampen the program. In a study of sterilization laws and their implementation, Philip Reilly found that "more than one half of all eugenic sterilizations [in the United States] occurred after the Nazi program was fully operational."[48] These laws had public support. According to a 1937 *Fortune* poll, 66 percent of the public favored involuntary sterilization of the mentally retarded, and 63 percent favored sterilization of habitual criminals.[49]

The early eugenics movement provided a way for those with means in society to avoid meeting the civil rights demands of the poor and the working class. Slaves were freed, but one of their basic civil rights was curtailed in the name of eugenics. African American citizens were thought to be so inferior that interracial marriage was prohibited to prevent the birth of defective offspring.[50] Starting in 1895, thirty-four states forbade such marriages by statute.[51]

Also in the latter part of the nineteenth century, there was growing labor unrest, culminating in the 1866 Haymarket bombing and riots.[52] The demonstrations were blamed on presumably genetically inferior immigrant workers and led to a series of immigration laws to keep people thought to be genetically undesirable from entering the country.

In 1882 a law was adopted to prohibit the immigration of people who were lunatics or idiots or who were likely to become public charges.[53] Later, in 1924, the U.S. Congress passed an immigration act setting quotas on the number of immigrants from various countries. This law was influenced by testimony that the U.S. gene pool was endangered by a large influx of people from southern and eastern European nations.[54] The inadequacy of the people trying to gain entry to the United States was "demonstrated" by a researcher who administered pen-and-paper IQ tests to exhausted, frightened individuals who had just landed on Ellis Island. The results: 87 percent of the Russians were found to be feebleminded, as were 83 percent of the Jews, 80 percent of the Hungarians, and 79 percent of the Italians.[55] Immigration was criticized as potentially making the American population "darker in pigmentation, smaller in stature, more mercurial . . . more given to crimes of larceny, kidnapping, assault, murder, rape and sex-immorality."[56] As vice president, Calvin Coolidge publicly declared: "America must be kept American. Biological law shows . . . that Nordics deteriorate when mixed with other races."[57]

The earlier science and law have been rebuked, but many of the policy questions raised a century ago have resurfaced in other guises. Legislators are debating whether people should be tested genetically without their consent. Social institutions are developing policies about whether it is proper to discriminate against people based on their genetic profile. In addressing the wide sweep of policy issues that genetic technologies raise, it is important to try to develop a contemporary framework for considering these dilemmas. It is also important to be aware of what type of political ideologies we are fostering with our policy decisions about the use of genetic technologies.

Alternative Frameworks

The diagnosis and treatment of disease have long been the focus of social policy debates. Concerns about access to health care services, their cost, quality, and social impact have led to a variety of professional, institutional, and legal guidelines—from medical licensing laws to drug safety regulations. Three different models dominate the regulation of health care services today: the medical model, the public health model, and the fundamental rights model. The application of one model rather than another depends on the perceived risks and benefits of the health care service at issue and the aspect of our lives that the service addresses. Choosing the wrong public policy response to genetics can aggravate negative impacts of genetic services. "The imprimatur of public policy can foster beliefs that eventually prove to have little basis in fact," observes geneticist Neil A. Holtzman.[58]

Canadian minister of health David Dingwall notes that the policies we develop are a crucial statement about who we are: "New genetic technologies concern the future of our society. How we manage them will be no less than a statement of who we are and what we value."[59]

Which model provides an appropriate starting place for genetics policy? Obviously, different genetic services will have features that require special policies. But it is useful to choose one of the three policy models as a starting point—as the place where policy analysts begin when crafting new policy— to simplify their task by not forcing them to create a central set of principles anew for each genetic development that appears on the scene. It will thus avoid what University of Wisconsin law and medicine professor Alta Charo refers to as the "bioethics fire drill," in which principles are developed in a panicked rush. Such a rush occurred when the cloning of a mammal was announced, and has also occurred, to a lesser extent, whenever a new gene is discovered.

The Medical Model

Currently, most genetic services are being offered under the *medical model*, the most common approach to setting policy for health care services in the United States. The medical model assumes that people will have access to particular health care services, so long as they can afford them and so long as health care providers are willing to offer them. Although this model emphasizes individual patient decisions, the physician is ultimately the gatekeeper for services and can be quite directive in recommending services. Issues such as quality assurance and confidentiality are generally left to the standards of the medical profession.

The underlying premise of the medical model is that physicians can judge which interventions a patient requires. In fact, for many generations, patients were expected to do whatever the doctor ordered. In 1957, for the first time, a California court introduced into the medical model a legal requirement of informed consent: a physician must give patients sufficient information in advance of a proposed treatment so that they are able to make a knowing decision about whether to undergo it.[60] At least twenty-three states adopted informed consent statutes requiring physicians to give patients certain information before a proposed treatment (and, in some instances, before a diagnostic procedure as well), including the nature of the patient's condition and the risks, benefits, and alternatives to the proposed intervention.[61]

The legal ideal of informed consent is rarely seen in practice, however.[62] A typical study found that "it was common in the hospitals studied for physicians to fail to inform patients about the nature, purpose, and risks of a planned procedure in a way that would enable them to make meaningful decisions."[63] In part, physicians' lack of disclosure is attributable to a profound misunderstanding of what patients actually want to know. While only 13 percent of physicians in one study said they would give "a straight statistical prognosis" to patients with advanced lung cancer, 85 percent of the public wished to have that sort of information.[64] A study on communication between physicians and patients with cancer showed that 75 percent of elderly patients diagnosed with cancer felt that their doctors "created undue worry by not providing them with enough information."[65] Another study found that 40 percent of cancer patients felt they were not fully informed about their diagnosis, prognosis, and treatment.[66] Moreover, in most states the scope of the disclosure is determined by the medical profession itself. Physicians are required only to disclose information that other physicians standardly disclose. In a minority of states, the standard is measured by patient needs; physicians in those states must disclose what a reasonable patient would want to know before making the decision.[67]

Thus, under the medical model, little attention is actually paid to informed consent. This is thought to be tolerable since people seek medical services when they already have a health problem and physicians are presumed to be acting in the patient's best interest by providing services. For example, there is little protest when certain routine, noninvasive blood tests are undertaken absent advance explanation.

Medical malpractice suits are the mechanism on which the medical model relies to address concerns about the quality of care. Patients may sue health care providers who fail to meet the "standard of care." Unlike other areas of law, where the standards of behavior are externally imposed, in medicine the standard of care is set by the profession itself. A physician has a duty to follow the standards set by the majority of the profession (or, at least, a "respectable minority"). Even if the profession provides abysmal care for a particular disorder, a physician who meets that low standard is not liable for bad outcomes. Only on extremely rare occasions have courts ruled that the standards of a particular medical field were so low that a different standard should govern.[68]

The medical malpractice system of quality assurance relies on a particular signaling method for error. It assumes that the patient is able to evaluate his or her worsened condition and decide that an error has occurred. When a physician fails to diagnose a cancer and the condition manifests, the patient is usually aware that an error has been made. When an improper medical or surgical treatment is undertaken, the patient may be aware of the error because of the resulting harm or because of the need for corrective treatment. Of course, this approach is not perfect. Most patients whose conditions worsen will believe it is the result of their underlying illness and not realize that it is the result of the physician's negligence.

Currently, most genetic services are regulated by the medical model. Under it, physicians are the source of information about genetic tests—though the fast pace of gene discovery may make it difficult for them to stay on top of developments. If genetic tests or services are performed negligently, the only recourse would be a malpractice suit—even though the harm might not be discovered until years later, which might exceed the time within which patients are allowed to bring suit. An erroneous "normal" result on a genetic cancer test, for example, may not be discovered until decades later when the person develops cancer, and certain side effects of gene therapy on embryos—such as sterility—might be impossible to detect until the resulting child reaches reproductive age.

The classic situation in which the medical model is appropriate is where physicians have a high level of knowledge about the health care service at issue, the service has a clear benefit, negligent provision of the service is easy to detect, and use of the service does not have potential to harm the patient physically, psychologically, or socially in a way that might cause the patient to reject the service. Determining whether the medical model should apply requires an assessment of the capability of the medical system to deliver genetic services in a high-quality way and of the extent to which new genetic services such as testing to predict later-onset disorders require more attention to patient consent than do other medical services.

The Public Health Model

A less common approach for regulating health care services is the *public health model*,[69] which attempts to prevent disease through education, financing of certain health care services, and, in some instances, mandated interventions (such as vaccinations). Generally, public health measures have been invoked to prevent imminent, substantial hazards to the population at large through efforts to eradicate infectious disease.

The classic mandate for public health occurs when health benefits can be realized only through intervention that requires participation of the public as a whole, such as the development of sewers, fluoridation of water, programs to screen for certain infectious diseases, and vaccination.

Mandatory intervention—such as vaccination—is used very rarely in health care. The laws that were adopted to require vaccinations—and the U.S. Supreme Court decision that upheld them—occurred at a time when infectious diseases such as smallpox and typhoid were threatening to kill off whole towns.[70] Smallpox had killed 60 million people in eighteenth-century Europe.[71] The court likened society's ability to vaccinate people to its ability to draft citizens to defend itself in wartime.[72]

The principle behind the vaccination laws is one of reciprocity[73] —mandatory vaccination of a particular individual protects other members of the community, while mandatory vaccination of others protects that individual. Historically, courts have not ordered nontherapeutic medical interventions for the benefit of an identifiable third party, rather than the community. For example, neither a cousin[74] nor a father[75] nor a half sibling[76] has been forced to provide bone marrow to a patient. Nor has a hospital been required to disclose the name of a potential bone marrow donor to a patient dying of leukemia.[77]

Another prime tenet of public health has been education. At the founding of the field, education focused mainly on information about the prevention of infectious disease. More recently, public health education has been directed at disease prevention through antismoking campaigns, proper nutrition messages, and advice for pregnant women, which all provide information to people who may not perceive themselves as ill and thus may not be seeing a physician who could otherwise provide that educational message.

There is currently a move to apply the public health model to an increasing number of genetic services. Genetics is called "the ultimate public health issue," by Muin Khoury, head of the Centers for Disease Control's new Office of Genetics.[78] The policies currently in place to address genetics within a public health model[79] include efforts to enhance awareness of and encourage the use of genetic technologies, such as a California regulation requiring obstetricians to offer maternal serum alpha-fetoprotein (MSAFP) testing to pregnant women in order to evaluate whether their fetuses have spina bifida and neural tube defects. They also include efforts to make genetic services available to low-income people, such as the public funding of amniocentesis under the Medicaid program in forty-five states.[80]

States have also adopted—and, in some cases, repealed—laws that required people to use certain genetic services. At the turn of the century, statutes in more than half the states required sterilization of institutionalized people who were thought to have disfavored genetic traits.[81] In the early 1970s, many states passed laws mandating sickle cell anemia screening of African American individuals.[82] Both of these sets of laws have been repealed. Currently, state laws mandate the use of genetic services in only one instance—newborn genetic screening. In five states, laws mandate that blood samples be taken of newborns to be tested for genetic disorders such as phenylketonuria (PKU) and congenital hypothyroidism.[83] In forty other states, parents ostensibly have the right to refuse newborn screening,[84] but because parents in most of those states are not told that they have such a right, testing is de facto mandatory there too. In the remaining states, newborn screening is voluntary.

The public health approach involves certain dangers. Interventions may be adopted before their risks have been adequately assessed. In the past, well-meaning genetics programs were adopted prematurely and caused unintended yet significant harm. In the late 1960s, state public health departments began mandatory screening of all infants for phenylketonuria, which is a genetic disorder that can cause mental retardation if the child is not put on a special low-phenylalanine diet shortly after birth. Because the program was implemented without adequate previous research or monitoring of the treated children, some infants who did not have PKU died or suffered irreversible damage when put on the special diet.[85]

Some previous public health uses of genetic services have improperly ignored the social and psychological impacts of genetic technologies. The laws establishing mandatory screening programs for sickle cell anemia carrier status in the 1970s did not provide adequate counseling or protections of confidentiality. The people identified by testing as being carriers of the sickle cell gene were stigmatized and discriminated against in insurance and employment.[86]

The classic case for the use of the educational component of the public health model occurs when widespread consensus holds that a particular lifestyle choice (smoking, unprotected sexual intercourse, lack of pregnancy care) is dangerous and information will change that behavior. The mandatory intervention aspect of the public health model is most appropriately applied to prevent the transmission of serious diseases to large numbers of people (as in the case of some infectious diseases). Determining whether genetics should be handled within the public health model requires an assessment not only of the seriousness of a disorder but also of whether prevention can adequately be achieved—and, indeed, whether prevention itself is an appropriate goal. While society might achieve near consensus about the appropriateness of vaccination to prevent measles, it is unlikely there would be much support for mandatory prenatal diagnosis and abortion as a way to prevent particular genetic diseases.

The Fundamental Rights Model

A third approach, the *fundamental rights model,* is applied to health care services that are central to our notions of ourselves, such as reproductive services. The fundamental rights model attempts to ensure that a health care service takes place only voluntarily, with extensive information provided in advance, and with quality assurance mechanisms in place. Greater justification is required under the fundamental rights model for governmental restriction on health care services—for example, abortion cannot be banned. The logic is that decisions about reproduction are important ways of expressing ourselves and that they have a vast impact on our lives. Decisions about whether or not to have children express an individual's personal beliefs and have a significant impact on his or her lifestyle. Certain U.S. Supreme Court decisions, for example, recognize that the ability of women to obtain contraception and abortion allows them to pursue other important means of personal development, such as education and employment.[87] Under the fundamental rights approach, use of a particular health care service (such as sterilization or abortion) must be strictly voluntary and uncoerced.[88] In addition, an individual is entitled to a great deal of information before undergoing the procedure. Under a federal law, for example, in vitro fertilization clinics must disclose their success rates.[89] In several states lengthy disclosures must be made about the medical and psychological background of proposed surrogate mothers.[90]

The fundamental rights model allows enhanced regulation for quality assurance when the usual tort incentives for behaving non-negligently are not operating with as great a force in a particular field as in other medical areas. This is particularly true with respect to reproductive technologies, where the harm from an error might not be discovered until the next generation. In addition, there is not the same signaling system about malpractice with respect to reproductive technologies as there is in other areas of medicine. When a patient is ill and undergoes a negligent surgery, the patient's condition worsens. But when a healthy couple undergo negligent in vitro fertilization, their own health does not generally suffer. Since in vitro fertilization has only about a 25 percent success rate, the couple may not realize (or be able to prove) that negligence has occurred. They may think (or be led to believe) that they are just in the unlucky 75 percent. Consequently, some states have adopted statutes to regulate the qualifications of the personnel who perform IVF[91] and to require that clinics disclose success rates.[92]

The fundamental rights approach also plays another role—protecting certain groups of patients from discrimination. The U.S. Constitution guarantees equal protection of laws; consequently government-funded entities cannot discriminate against patients based on race, gender, religion, or ethnic status. Additional federal and state statutes provide further protection. The federal Americans with Disabilities Act, for example, prohibits employers, health care providers, and other groups from discriminating against patients based on their disability.[93] Some state human rights statutes prohibit discrimination against people based on race, gender, and marital status.[94] City ordinances, too, may provide additional protections, such as prohibition against discrimination based on sexual preference.[95] In some instances, the fundamental rights approach also serves as a justification for requiring public funding for additional genetic services for people who cannot afford them.[96]

The fundamental rights approach has been applied to certain genetic services—most often, those involving reproduction. A federal court case, *Lifchez v. Hartigan*,[97] held that since people have a fundamental right to privacy to make reproductive decisions, they have a fundamental right to information upon which to make those decisions. Consequently the court struck down a state statute banning embryo and fetal research because it prohibited couples from using experimental forms of prenatal testing (such as chorionic villi sampling, which was considered experimental at the time) to learn genetic information about the fetus. A parallel action has been taken by legislatures in six states which contain exceptions in their bans on embryo and fetal research to allow experimental genetic screening of embryos.[98]

Under the fundamental rights approach, people are entitled to more information about medical services than under the medical model. This approach has already been invoked in the genetic context in court cases dealing with physicians' failure to advise pregnant women over age thirty-five of the availability of prenatal testing, such as amniocentesis (in cases where the women have subsequently given birth to an affected child).[99] In those cases the physicians were held liable for not informing the women even though, under a medical model, the medical standard of care at the time did not require that they be informed (since most physicians did not do so).

The fundamental rights approach has also been used by state legislatures to prevent discrimination against people based on their genetic status. Some states, for example, prohibit the use of genetic information in employment decisions or insurance decisions.[100]

The fundamental rights approach is used for decisions about which health care providers themselves may have an inadequate understanding or may be unduly influenced by their personal feelings. It is considered to be the most appropriate approach when the decision to use or refuse a particular health care service has an impact on how the individual is viewed and treated by social institutions. It also comes into play if traditional malpractice law is inadequate to assure quality.

Seeking a Framework for the Future of Genetics

The lack of systematic policymaking guidelines regarding new genetic technologies in the United States, compounded with the search for financial gain from genetic innovations in our society, has led to inefficient and ineffectual regulations and laws concerning the use of genetic services. In order to avoid the "bioethics fire drill" to which Alta Charo refers, a basic policymaking framework coupled with appropriate oversight is necessary if we are to tackle the issues in an organized manner and not in the narrow, piece-by-piece manner of today. Given the specter of eugenics that hangs over our society we must tread carefully when making policy decisions.

Which approach is the best starting point for designing policies for ge- netics—the medical model, the public health model, or the fundamental rights model? The answer to that question must be based on an analysis of the potential impact of each approach, taking into account the effect of particular genetic services on an individual's self-concept, relationships with family members, and relationships with social institutions. It requires an analysis of the current level of quality of genetic services and the ability of health care professionals and institutions to provide those services. It also necessitates a close scrutiny of how genetic services may disproportionately affect certain groups within society, such as women, people of color, and people with disabilities. Finally, it requires an analysis of the social context in which genetic services take place and the likely impact of those services on fundamental cultural concepts.

3 The Impact of Genetic Services on Personal Life

Genetic testing generates information unparalleled in other areas of medicine. People can learn that, decades later, they will suffer from an untreatable disorder. People can learn that they have an increased risk of cancer—or that their children have a one-in-four chance of dying of a serious disorder in childhood. The impact of this profound knowledge ripples throughout people's lives—challenging their self-image, changing their relationships with family and friends, and causing them to think about their life, health, and responsibilities in new ways.

Impact on Self-Concept

Learning predictive genetic information about themselves can cause people to view themselves differently. Genetic testing—offering a glimpse into a person's most intimate self, as well as assessment about potential risks in his or her medical future—can have a profound influence on his or her personal life. It can affect people's self-images and relationships with those closest to them. It can change their ideas about health and their relationship with the medical system.

Decisions about whether to use genetic technologies—and the impact of their use—can threaten an individual's psychological well-being and self-concept.[1] People come to genetic technologies with a particular sense of self, sense of morality, and sense of responsibilities. Their response to genetics can cause them to reevaluate their view of their personality, their self-worth, their sense of security, and their relationships with loved ones.

Anxiety and Distress

Learning genetic information about oneself or one's fetus can cause anxiety and distress. People who learn they are carriers of a gene mutation for a recessive disorder may find that their view of themselves is changed by this information. Generally the presence of such a gene mutation becomes a problem only if the person has a child with someone else who has that same mutation. Then there is, at most, a one-in-four chance that the child will be affected. Yet some people who learn that they are carriers describe themselves as "marked" or "unmarriageable." Being identified as a carrier of a recessive genetic disease generates anxiety, which may have a lasting impact on the individual.[2] In a cystic fibrosis carrier screening study, 27 percent of carriers remained anxious about test results six months after testing.[3] Similarly, in an eight-year follow-up of individuals who had been screened for Tay-Sachs carrier status in high school, 46 percent recalled that they were upset when they received the test results, and 19 percent remained worried eight years later.[4]

Carrier status has no effect on the individual's physical health, and although he or she may understand that fact, carriers as a whole have more negative feelings about their future health than members of the general population.[5] When carriers are asked about their future health they describe it in gloomier terms than noncarriers. (At a deep subconscious level, they may feel that if this "bad" thing has happened to them, they are at risk for other bad things as well.) Similarly, people who are at risk of a genetic disorder because their parents have manifested a disease have gloomier ideas about the future. A study of the daughters of women with breast cancer found that the daughters had less confidence that their internal and external environments would be predictable and that things would most likely work out as well as could be expected.[6]

Both men and women may feel inferior if they learn they are carriers of a recessive genetic disorder. In Sardinia, for example, there is a high incidence of beta thalassemia. Extensive education is undertaken in the schools to explain that this is a recessive disorder in which both members of the couple must carry a mutation in the gene for a child to be affected. Nevertheless, when an affected child is born, says one Sardinian genetic counselor, it is not uncommon for each parent to point to the other and say, "This disease must be from your side of the family."

Presymptomatic genetic testing for late-onset disorders can be even more problematic, since the results may signal future health risks for an individual. In a preliminary study of BRCA1 testing for a predisposition for breast cancer, more women with the mutation experienced psychological distress than women without the mutation.[7] Caryn Lerman and Robert Croyle urge that the psychological consequences of DNA test results be considered to be at least as significant as the impact of learning that one is at risk due to family history. As to that more general risk, 27 percent of high-risk women who self-referred to a breast cancer screening project had symptoms warranting psychological counseling.[8]

People's life choices may be affected by their perception of genetic risk, yet those perceptions might not be correct.[9] When considering the likelihood of having an altered cancer susceptibility gene, women from high-risk breast and ovarian cancer families report that they believe that having a mutation is nearly certain.[10] When the risk estimates made by women in high-risk groups were compared to a mathematical model, it became clear that the women were overestimating their cancer risks: 43 percent rated their risk of having a BRCA1 or BRCA2 mutation as between 76 and 100 percent, but the model estimated that only 26 percent had that high a risk of having the mutation.[11] The average difference between the women's estimates and the model estimate was 32 percent.[12]

The nature of the disease at issue affects distress levels. Individuals at risk for Huntington's disease have higher anxiety levels than those at risk for colon cancer.[13] Those at risk for neurological diseases are generally more distressed than those at risk for cancer. This may be related to their fear of a loss of their mental "self" and the lack of preventive measures.[14]

Previous experience with a disorder can also increase distress levels and the expected impact of genetic testing information. Specifically, people who saw their "father tied to his chair" or laughed at for the drunken-like behavior characteristic of Huntington's disease had higher levels of distress than those who had had no experience with the disease.[15]

Individuals who know they are at risk for a particular disease because they have parents or other family members who carry a faulty gene may make conscious or subconscious choices about their lives based on their assumptions about whether or not they will get the disease.[16] Receiving the results of presymptomatic testing—regardless of whether the results indicate future disease—can be psychologically troubling because it can run counter to a

person's expectations. A woman might have deliberately refrained from having children in order to avoid the possibility of passing on a dominant disease
that affected one of her parents. Years later, she may experience strong regrets
about this decision when she then learns she does not have the mutation,
and thus her children would not have been afflicted with the disease.

The person who learns she has—or is likely to get—a genetic disease may
change her view of herself. Hollis Sigler, an artist with breast cancer, has
painted a series of works titled *Breast Cancer Journal: Walking with the
Ghosts of My Grandmother*. Along the frame of one of the paintings, she
wrote: "The guilt of responsibility—could I have wished this on myself? In
this culture, which places so much emphasis on the individual, it is seen as
a personal failing if one has a disease. What did I do that was wrong? Can
I make it right again?"[17] Women who learn they have a breast cancer gene
mutation often feel differently about their bodies. After DNA testing revealed
she had a breast cancer gene mutation, one woman said, "It felt as if there
was a time bomb ticking away inside me."[18]

The psychological impact of learning that one has the gene for an untreatable disorder such as Huntington's disease can be significant. Huntington's disease is a dominant disorder; the children of at-risk individuals have
a 50 percent chance of inheriting the disease. The disease generally does
not begin to affect people until after age forty. It progressively destroys the
brain's neurons, retarding motor skills and slowly causing confusion, irrita-
bility, depression, dementia, movement disorders, and death.[19] "Suicide is
now a question of when, not if," one woman wrote to a psychologist after
learning that she had the genetic mutation associated with Huntington's
disease. The suicide rate is four times higher among people with Huntington's disease than among the general population.[20] Even when people are
prepared for bad news as a result of testing, they may still be shocked by the
reality of it. A woman who expected that she had a faulty Huntington's
disease gene nonetheless stated, after she received the results, "I feel like
someone has died. Part of me has died, the hopeful part." The woman
experienced depression, which became increasingly problematic.[21]

A person's social life and home life influence his or her response to test
results. Among people who underwent genetic testing for colorectal cancer,
those who had lower levels of social support suffered more anxiety and de-
pressive symptoms than those who had more social support.[22] Supportive
partners can help an individual feel less helpless after receiving the results
of genetic testing.[23] The negative effects of learning genetic information are

worsened when the risk is high, when uncertainty is not reduced, and when the results are at odds with actions already taken (such as when a woman has forgone having a child because of concern that she would pass on the disease and then later learns she does not have the mutation).[24]

In some studies of cystic fibrosis testing, people from at-risk families testing negative and those testing positive did not differ in their feelings of anxiety,[25] but people from the general population who underwent testing and found that they were carriers were more anxious than they had been before testing.[26] Thus, people who have not previously been concerned about genetic disease may be particularly distressed by new knowledge about genetic mutations that might affect their reproductive plans. A similar response has occurred with regard to breast cancer; women who had never experienced breast cancer themselves were more distressed when they learned they had a BRCA1 mutation than women with previous cancer: "Because the vast majority of women in population-based screening programs would not have had cancer or cancer-related operations," write Robert Croyle and his colleagues at the University of Utah, "our finding that distress is highest among carriers with these characteristics has significant public health implications."[27]

Impact of a Negative Test Result

Monica Bradlee lived for twenty years with the fear that someday she would discover she had Huntington's disease and would suffer the slow debilitation that she had seen her father go through. She also worried about the risk for her two children. Believing that "the devil you know is far better than the one you don't know," Monica was tested. Initially elated upon hearing that she did not have the gene, Monica then became "oddly confused, unsettled, and not, as one might have imagined, at peace." Monica said, "It's taken a long time for me to be able to feel the good news. . . . I don't know who I am anymore or what my goals are. The whole world is open to me now." She feels guilty and often wonders, "Why was I the lucky one?"[28]

While testing negative for a genetic mutation can provide a tremendous sense of relief to an individual, that is not always the case.[29] Even when the results of genetic testing reveal that a person does *not* have a genetic mutation, the results may nevertheless cause psychological harm to him or her.

Some people experience "survivor's guilt," similar to that of soldiers whose buddies have died in war, as they wonder why they have been spared when other family members tragically have inherited the gene.[30]

Because more people have undergone predictive testing for Huntington's disease than any other adult-onset disease, it has served as a touchstone for how to use genetic tests and counsel patients.[31] According to Barbara Bowles Biesecker, genetic counselor at the Human Genome Research Institute, many important lessons can be learned from this example. "The psychological issues for people who find they don't have the gene are often as dramatic, or more dramatic, than for those who carry the gene," she says. She suggests that all individuals be given their test results in person in the company of someone who can provide them with support and that they have at least one follow-up counseling session. In addition to "survivor's guilt," individuals who thought that they were going to get the disease may suffer an identity crisis from a negative test result, as did Monica Bradlee.

Of people who undergo genetic testing for Huntington's disease and learn that they do not have the gene, 10 percent experience severe psychological problems as a result.[32] Many people who have a parent with Huntington's disease assume they have inherited the gene.[33] They may live their lives as if they will die of the disease in their fifties. They may have chosen not to pursue a particular career or relationship because they believe they will get the disease. Learning they do not have the gene radically changes their self-image. One woman said, "If I'm not at risk—Who am I?"[34]

One man who was at 50 percent risk for Huntington's disease because one of his parents had it lived his life in preparation for an early death. He enjoyed dangerous hobbies such as skydiving. He spent money rather than saving it and ran up huge loans and credit card bills. He did not commit to a long-term relationship with a girlfriend because he did not want to get married and risk passing on the gene to a child. Then he was tested and learned he did not have the gene associated with Huntington's disease. He was just an ordinary Joe, who might need to start thinking about a mortgage rather than living hard, dying young, and leaving a beautiful corpse. The test results precipitated a downward spiral, and he embezzled from his company to pay his bills.[35]

This case, and many like it, show how the identity a person brings to genetic testing influences his or her response to the results. In at-risk families, some members assume they will have the mutation. A person with Aunt Betty's red hair and sense of humor might feel that she will also get Aunt Betty's breast cancer or Huntington's disease as well.[36]

Moreover, people without the genetic mutation for Huntington's disease often feel some initial relief at the results, but three years later, they tend to feel as helpless as do carriers of the mutation. "Presumably, the test result has not brought the expected effect on their life perspective," note Dutch researcher Aad Tibben and his colleagues at Erasmus University. "Contrary to their expectations, the favorable result has apparently not conferred new mental resources to make decisions or solve personal problems."[37] Furthermore, the process of genetic testing can cause the individual to relive the trauma of caring for and watching the demise of an affected relative. In one study, "noncarriers reported that the result had activated the past and that they were overwhelmed by early experiences with Huntington's disease in the affected parent."[38] In a Norwegian study, half the people undergoing presymptomatic testing for Huntington's disease needed psychiatric treatment during the process. One-quarter said they had wanted a closer follow-up after getting the results. The majority of those wanting more follow-up had tested negative.[39]

A negative result could relieve an individual and renew his or her hope of living a healthy life, but testing negative for certain mutations does not necessarily mean that one will not get the disease. For example, heredity is thought to account for only 5 to 10 percent of all cancer cases.[40]

Genetic Information and Identity

Genetic information provides more than potential health information — it provides a new way of looking at oneself. Nowhere is that more clear than in Iota, Louisiana, a place where, in the past, being Jewish was considered something to be ashamed of. The New Orleans historical archives contend that no Jewish person set foot in New Orleans until the late nineteenth century.

In actuality, according to a ship's manifest, a German Jew, Johann Adam Edelmeirer, crossed the ocean and landed in Louisiana in 1720. It is now assumed that he brought with him more than his worldly goods. He brought with him the genetic mutation, common to 1 percent of Ashkenazi Jews, for the devastating disorder known as Tay-Sachs disease.

Once in Louisiana, Edelmeirer probably hid his Jewish origins because of a 1724 regulation in the territory requiring that all Jews leave within three months upon penalty of confiscation of their persons and property.[41] These facts might not have been unearthed were it not for the recent advent of genetic testing. Just a few years ago, Cajun Catholic and fundamentalist families who were descendants of Edelmeirer learned that they carried the Tay-Sachs mutation, a "Jewish" mutation.[42]

The psychological and social upheaval caused by this information in such a family can be enormous, yet these scenarios will be repeated as individuals are tested and receive genetic information that appears linked to an ethnic and racial background about which they were unaware.[43] In addition to the psychological changes in self-image, there may be dramatic changes in social standing and even in finances. For example, a person who is socially considered a Native American yet who does not meet what researchers claim is the typical genetic profile of Native Americans may in the future be deprived of certain land rights and educational scholarships.[44]

In Orthodox Judaism, the most respected and highest-status rabbis are thought to descend from Aaron, the older brother of Moses, who was the first of the Cohanim, the Jewish priesthood that predated the rabbis. In Orthodox and some Conservative congregations, these descendants are accorded a particular high status and are the only rabbis who can perform certain religious duties. Now researchers have found a genetic pattern on the Y chromosome that they believe is shared by the descendants of the Cohanim.[45] Could such a test be used to remove rabbis from the high priesthood if their DNA does not measure up?

"There are no Jewish genes. There are no black genes," says Dr. Richard Klausner, director of the National Cancer Institute.[46] Everyone has between five and fifty genetic mutations. No individual or race is perfect. Despite these facts, associations of genetic diseases within specific cultures have instilled fear of discrimination and stigmatization among those groups. Individuals in many ethnic and racial communities have been identified as higher-risk candidates for certain diseases.

The Jewish community is concerned about being perceived as genetically inferior in light of the Holocaust. This has led some Jewish individuals to refuse to participate in genetic research, in order not to have their group targeted as inferior. Such concerns left Boston researchers, who were hoping to study breast cancer, without subjects. In contrast, in Washington, D.C., where researchers worked closely with Jewish leaders to explain the study and its benefits and to assuage fears about privacy, more than five thousand Jewish women were willing to participate.

Genetic tests that purport to predict certain behaviors can also profoundly affect self-image. What will be the impact on a married man who believes himself to be heterosexual yet learns he has a gene that is purportedly linked to male sexual preference?[47] What about those individuals who learn they have a gene alleged to be linked to aggression?[48]

Nicola Barry noticed that her "perfect baby boy, intelligent, lively, and handsome" was slow in some areas. At age seventeen he was diagnosed with Fragile X, named after a faulty X chromosome. Nicola found herself defending her son, and children like him, after a story ran in newspapers with the headline HORROR STORY IN THE X FILES. The story identified the "X gene as the 'delinquency gene.'" It recommended that screening for the disease could, as Nicola paraphrased, "make prison a thing of the past by identifying the rogue gene before birth and aborting the babies who would otherwise grow up to be future criminals." No studies have linked Fragile X to delinquency. Nicola remarked, "I wasn't just insulted, I was worried as well because misreporting on that sort of scale is highly dangerous. . . . It created all kinds of problems for families with affected children."[49]

Indirect Information on Genetic Status

An individual does not just get genetic information from his or her own testing. The testing of relatives also provides some risk-status information about an individual, and the testing of a fetus can provide definitive information about the parents.

Genetic testing of a fetus can have a tremendous impact on the parents' emotional well-being and self-concept.[50] If the fetus is affected with a recessive disorder, that means that both parents are carriers of the gene for the disorder, though they themselves do not have the disorder. In addition, the information—which often the parents would not have found out if the mother had not undergone prenatal testing in conjunction with reproduction—may change their views of themselves, making them feel "defective" when they risk producing what they or others view as an "abnormal" child.

If a fetus is being tested for a dominant disorder, diagnosis of a genetic mutation means that the parent who is at risk has it as well. This comes up most strikingly with respect to the debilitating neurological disorder Huntington's disease, which does not generally begin to affect people until after age forty. Most people who have a parent with the disorder, and thus are at 50 percent risk of developing the disorder themselves, decide not to get tested. They see no benefit in learning their status, since no treatment is available. Thus, before the fetus is tested, the at-risk parent usually does not know whether or not he or she will later develop the disorder. But testing the fetus also discloses information about the parent's genes. Psychologist Nancy Wexler observes that if the test of the fetus for Huntington's disease is positive, "two massive losses are suffered simultaneously. The child will most likely be aborted, or else the test would not have been taken initially. And the parent is immediately diagnosed as an obligate carrier." If the disorder is a dominant late-onset untreatable disorder such as Huntington's disease, at-risk parents "hear their own death knell with that of their child."[51]

Use of Genetic Services

At the University Hospital at Stony Brook, two women considering whether or not to be tested for mutations of the BRCA1 gene consulted the same genetic counselor, Laura Lesch. Lesch explained to both women that even if the test is accurate, regardless of the results, it cannot definitively predict the occurrence of breast cancer. One woman, who lost a mother, sister, and two aunts to the disease, decided to submit blood for genetic testing. She paid for the test herself because she was concerned about how her insurance company might respond if they found out she carried a mutation of BRCA1. The second woman, who had heard the same information from Lesch, decided she would prefer not knowing to risking the loss of her insurance coverage or her job, especially since the test could not accurately predict whether or not she would get breast cancer.[52]

Based on their own personalities, backgrounds, and life contexts, individuals may react in a variety of ways to the offer of genetic testing and the information generated by a test. Some geneticists fail to appreciate the many complex factors that influence a person's response to genetics and assume that people will invariably find it beneficial to learn genetic information, whether it brings good news or ill.

That kind of assumption is evident in predictions about uptake of genetic services. Cystic fibrosis (CF) is the most common potentially fatal genetic disorder in whites. Those who have one gene with a CF mutation are unaffected carriers of the recessive disorder. If two carriers produce a child together, there is a 25 percent chance that the child will be affected with cystic fibrosis, a disorder of the exocrine glands that causes chronic obstructive lung disease.[53] Many such children need treatment, but now live to an average age of thirty;[54] some individuals lead normal lives without realizing that they have two genes with the mutation.

When the cystic fibrosis gene was discovered, the assumption was that virtually all whites of reproductive age would rush out and be tested. In some analyses, it was suggested that once testing became available, each genetics health care provider in the United States would spend at least sixteen weeks per year counseling individuals about cystic fibrosis testing.[55]

In an initial survey asking patients in obstetrics and gynecology clinics if they would be interested in CF testing once it became available, 84 percent said yes.[56] But when a series of pilot studies was introduced in which CF testing was offered free of charge to the general population, very few people wanted the test. In a Vanderbilt study, fewer than 1 percent of people offered the test took it.[57] In a Johns Hopkins study, an HMO mail campaign offering free cystic fibrosis carrier testing to patients garnered a 4 percent response. When patients already at the HMO were approached for free CF carrier status screening, only 24 percent chose to undergo testing.[58]

With respect to testing for late-onset disorders, geneticists also expected widespread interest. When a genetic marker for Huntington's disease was identified in 1983,[59] with presymptomatic marker testing in 1986,[60] and, later, when the gene itself was localized in 1993,[61] it was expected that at-risk individuals would flock to testing. Geneticists assumed there would be a benefit to individuals who received HD testing results.[62] They thought that those who tested negative (and thus did not have the gene associated with Huntington's disease) would be overjoyed at the results. They also assumed that those who tested positive—and learned they would develop Huntington's disease—would, at some level, value the information since it would allow them to purchase insurance or otherwise make financial provisions for their future illness or make lifestyle changes, such as refraining from having children.

The initial surveys of at-risk individuals seemed to substantiate geneticists' beliefs. The majority of at-risk individuals said they would undergo HD testing if a test were to become available.[63] When the test actually became available, however, fewer than 15 percent of at-risk individuals chose to undergo the testing.[64] Similarly, many people with relatives affected with breast cancer said that they would undergo genetic testing if it were to become available,[65] but once they learned more about the uncertainties and risks of testing, fewer women than expected pursued such testing.[66]

The low uptake on testing is one indication of the complexities of the decision to use genetic technologies. The reactions that people have to their test results further indicate that genetic technologies are not an unalloyed benefit. Though scientists assume that if the public is educated about new genetic technologies, they will be more supportive of them, this is often not the case. The more people know about basic biology, the more likely it is for their optimism about the potential contribution of biotechnology and genetic engineering to decrease. The more women know about the limited predictive value of breast cancer testing, the less likely they are to be interested in having the test.[67]

Impact on Preventive Activities

Genetic testing is often thought to be beneficial even for people who test positive as a way to motivate them to undertake more health surveillance or prevention activities and to help them cope better if the disease does later manifest. Various studies, however, have found that these benefits do not necessarily occur after genetic testing. In fact, there is evidence that the stress created by genetic information can actually lessen the likelihood that the individual will engage in surveillance strategies and monitor himself or herself for early signs of the disease.[68] Even if the person does engage in preventive strategies, the physiological and psychological harm from increased stress may offset any benefit of increased surveillance.[69]

Nor does knowing one's risk prior to onset necessarily make the individual more able to cope with the disease once it manifests. Aad Tibben and colleagues at the University of Erasmus found that carriers who coped with the initial results of Huntington's disease testing became depressed, exhibited suicidal tendencies, showed disturbed functioning, or all of these once they manifested symptoms.[70]

Moreover, even when a treatment is offered to a presymptomatic individual, that treatment—such as "prophylactic" surgery based on presymptomatic testing—brings its own physical and psychological risks. The first woman with the BRCA1 breast cancer gene mutation who underwent a prophylactic mastectomy said, "I had wonderful counselling beforehand but nothing prepared me for the feeling of loss. There is a kind of feeling of mutilation. A woman's breasts are very much tied up with the image she has of herself and however perfect the reconstruction you are aware they are not your own. It was far more emotionally traumatic than I had expected."[71]

Genetic Services and the Psyche of the Individual

Overall, the studies about the impact of genetic information show a profound effect on individuals. People's most intimate sense of themselves and feelings of security can be shaken by the use of genetic services. They may gain information about themselves that does not comport with their previous view of themselves, that causes anxiety, and that makes them more worried about their health but, in certain instances, makes them less apt to do anything about it. Their use of genetic services may itself conflict with their previous views of themselves. In some instances, use of genetic services may violate religious tenets (such as when a Catholic woman aborts a fetus affected with a genetic disorder), or it may make individuals think less of themselves (such as when they abort a fetus with the same disorder as an existing child).

The existence of genetic tests can even have an influence on people who do not take the tests by making them feel responsible or guilty for bringing a child into the world without prenatal testing. Some geneticists and oncologists describe women who undergo breast cancer testing as "courageous." "I wasn't afraid," Dr. Rudolph Tanzi, an Alzheimer's researcher, told the *New York Times* about his decision to test himself for the genetic mutation purportedly associated with Alzheimer's disease. Some genetic counselors describe people who do not partake of genetic testing as "avoiders."[72] Such a characterization may cause people who refuse genetic testing to think less of themselves—or it may subtly coerce them into undergoing testing they do not want.

In part because of the role their physicians play, people tend to associate testing with treatment even when testing may not lead to a preventive treatment. Nancy Wexler points out, "People are told, 'If you're worried about getting it, go take a test.' That's the American way."[73]

The choice to use genetic services or not involves more than a bit of tissue, a sliver of DNA. It involves the whole person, with impacts that touch all aspects of that individual's life.

Impact of Genetic Services on Relationships with Family Members

Two identical twin brothers, both air traffic controllers, have a parent with Huntington's disease, and thus each has a 50 percent risk of carrying the gene and, hence, getting the disease. The first twin wants to learn if he has the gene. His brother does not. But even if the first brother vows he will keep his test results secret (since if he has the gene, his twin does as well), will he actually be able to do so? If he doesn't have the gene, won't he want to tell his brother the joyous news? And if he doesn't relay good news to his sibling, won't his brother then know that they both have inherited the gene? If the first brother's medical record indicates that he has the Huntington's disease gene, the second, untested brother could be denied health and life insurance based on those results—and both could lose their jobs as air traffic controllers.

The fact that we are inextricably bound to our kin by shared genes reaches far beyond the case of identical twins. A parent and a child have half their genes in common, as do siblings. Cousins share one-quarter of their genes, as do grandparents and grandchildren. In one instance, a forty-five-year-old woman decided not to be tested for Huntington's disease. But her twenty-five-year-old daughter, who was considering having a child, wanted to be tested to determine whether she had the gene. If the daughter did have the gene, her mother would thereby get the unwanted news that she herself had the gene and might soon be afflicted with Huntington's. Does the daughter have an obligation to accede to her mother's wishes and forgo testing?

Genetic services affect personal relationships in a variety of ways. The existence of genetic information within a family may create—or destroy—bonds between relatives. Genetic information makes some parents feel guilty about passing on a genetic mutation to their children. For a variety of common disorders such as breast cancer or heart disease, which were not until recently viewed as having a predictable heritable component, the fact that a test can now pinpoint the mutation that the parent has passed on makes parents feel more culpable for their children's illness.

Genetic testing within families creates new sources of tension and conflict. In some research studies, names and addresses of relatives are obtained from the patient.[74] These family members may be upset that their privacy was invaded. Some patients may not want to provide access to relatives, and researchers may pressure them for disclosure in order to recruit more participants. One grandmother with a genetic disorder was called repeatedly by researchers who wanted her to provide them access to her adult grandchildren so that the researchers could identify whether the family's particular mutation was traceable back to a certain ancestor in Europe. The repeated calls made her feel guilty about passing on a "bad" gene. They also made her worry that the researchers would coerce her grandchildren and that they might lose their health insurance as a result. "I wish I'd never gone to the doctors for treatment," she confessed to her granddaughter.

Within families, there may be pressure on relatives to be tested. For some genetic testing, linkage studies are still required, where blood relatives must provide tissue samples for a determination of whether a particular mutant gene (such as a particular form of the breast cancer gene) runs in the family. Only by learning about that family's form of the mutation can a definitive diagnosis in a particular individual be made. With breast cancer, for example, there are hundreds of ways in which the BRCA1 gene can vary, and clinicians must narrow down the particular distinctions for which to test. In order to use genetic linkage studies, the person who wants genetic information must persuade other members of the family to provide DNA for testing.

In one family, a woman had been raising a son with a neuromuscular disorder for years. She had been ostracized by her sister, whose own children were "normal." The sister never offered to watch the affected child to give the mother a day off. Then, when the sister's own healthy daughter got pregnant, she asked that her sick nephew submit to a blood test so that a linkage study could assess whether the fetus was at risk.

The response from the boy's mother was "Hell, no."

Should relatives be considered to have a moral (or even legal) duty to share information with each other — and provide blood samples? While some geneticists believe that genetic information should remain within an individual's private control, others refer to it as "family property." Dorothy Wertz and John Fletcher suggest, "It's vital to recognize that hereditary information is a family possession rather than simply a personal one." Other commentators question, "Why should a mere biological link justify an encroachment on an individual's sphere, in which most relatives are not socially or psychologically involved?"[75] The medical community is currently divided on the issue,[76] yet an individual seeking genetic testing may not realize that her choice of doctors could affect who will receive the results of her genetic testing without her consent.

Even short of situations in which a family member might be asked to undergo genetic testing, questions arise as to whether and when relatives have a duty to share genetic information. If a woman learns that she has a mutation of the breast cancer gene, does she have a responsibility to tell her sister that she, too, might be at risk? If she decides not to disclose, may her physician breach confidentiality and warn her sister? If a person has the genetic condition malignant hyperthermia (which can cause death if the individual is exposed to certain types of anesthesia), his or her sibling may have the condition as well and need to avoid those anesthetics. In other cases, however, the benefit of the genetic information is less clear. If a woman has the genetic mutation purportedly predisposing to breast cancer, her sister may or may not want to know that she, too, is at risk. Thus far, there have been no randomized clinical trials indicating that greater monitoring of women with the mutation and/or prophylactic mastectomy for women with the mutation improves the quality of life, psychological well-being, or medical outcome of those women.[77] And while some individuals who are at risk for Huntington's disease may wish to change their financial planning as a result of that information, others may not wish to know that they have a 50 percent risk of developing an untreatable disease. In some families, individuals may withdraw from family activities and events in order to avoid having to face family members with whom they do not wish to share test results.[78]

Complex emotional issues often engage the family as a whole, and family members separately, when predictive testing is underway.[79] In a study with the first family in the Netherlands for whom predictive DNA testing for hereditary breast and ovarian cancer became an option, two important roles within the family were observed: the messenger of the news and the first utilizer.[80]

The messenger of the news often is the one who motivates his or her family members to participate in a linkage study. The messenger may want to "rescue" other family members; however, he or she may have feelings of guilt, or other family members may direct their aggression and revulsion about predictive testing toward the messenger.[81]

The first family member to be an identified gene carrier in the family and to choose preventive surgery is called the first utilizer. This person is often under pressure to provide benefits for himself or herself as well as for the family. The family may not have any experience with predictive DNA testing; therefore, the first utilizer feels obliged to help the other members understand how the process works but may also feel an unexpected burden resulting from functioning as an "example" in the family.[82]

The web of relationships around genetic information runs in various directions. Children may want genetic information about their parents in order to be able to make their own reproductive plans. Parents may want genetic information about their young children in order to make medical and financial decisions about their children's futures or to prepare themselves psychologically as well. In every instance, the acquisition of genetic information has familial and social implications. A child may be denied health insurance later in life as a result of a genetic test requested by the parents when the child was very young.

The ripple effect of genetic information spreads beyond the nuclear family. When a pregnant woman learns genetic information about her fetus, that information indicates certain risks to relatives as well. If the information indicates that her fetus has cystic fibrosis, there is a 12.5 percent risk that her cousin is a carrier as well.

After discovering that Pru and Richard Irvine's son Henry had Fragile X, genetic counselors advised the couple to inform their siblings and cousins about the disorder and about the possibility of testing. Pru wondered, "Would anyone not want to be tested? Who would really choose to be emotionally battered and deprived of sleep? Who would really choose to take responsibility of someone who could not make sense of their world?" Pru herself had undergone sterilization. She wrote, "I could not take the smallest risk of having another baby."[83] Other people, however, look at the variable manifestation of Fragile X and find it appalling *not* to consider having an affected child.

Michele Olson, currently a waitress in Encinitas, California, was diagnosed with cystic fibrosis at the age of three. At twenty-nine, she was tested to determine her specific gene mutation in case a therapy would be available for it. Two years later, Michele sent a report of her genetic test to her cousins. They had never asked to receive the information. For them, the report of Michele's genetic mutation, whether they chose to save it or toss it, is, as one journalist put it, a "bewildering talisman of the future."[84] Some surveys have shown that individuals such as Michele's cousins, who are not themselves affected by a recessive genetic disease, would rather not know if they carry one copy of the genetic mutation.[85] Yet other people—and certain social institutions—may want to force them to learn that information since, if they have a child with another carrier, there will be a 25 percent chance that the child will inherit the disease.

Even if an individual decides not to share the genetic information with relatives so that they will not face psychological, social, and financial risks, they may learn about it anyway. There is a widespread willingness on the part of geneticists to breach a patient's confidentiality and disclose information to relatives. In response to a survey by Dorothy Wertz, a social scientist at the Shriver Center on Mental Retardation, and John Fletcher, a bioethicist at the University of Virginia, 60 percent of geneticists said they would disclose the risk of Huntington's disease to a relative without the patient's permission.[86] This is despite the fact that Huntington's disease is untreatable and that fewer than 15 percent of individuals who are at risk for it decide to undergo testing to determine whether they have the disease.[87]

Some people who are reluctant to learn genetic information about themselves seek to authorize analyses of tissue of their dead parents. Dr. Fiona Crawford, an Alzheimer's disease researcher at the University of South Florida, says that people who don't necessarily want their own APOE analyzed as a genetic test for Alzheimer's disease may want to know if their deceased parents had an APOE mutation.[88] If their parent did not, then they will not. If their parent did have a mutation, they can still tell themselves there is a 50 percent chance they did not inherit it (rather than face definitive information that might be revealed by testing themselves).

Genetic testing creates new decisions for families as a whole. In one instance, a couple learned the woman was carrying a fetus affected with cystic fibrosis.[89] They called together their family members and asked what type of emotional support they would get if the child were born. One group of relatives resented the decision, worrying that the couple would claim an excessive portion of the grandparents' time.

Genetic information can also change the dynamics within families. Psychologist Kimberly Quaid of Indiana University has seen how test results of siblings who each have the same chances of inheriting diseases from their parents have altered their relationships with each other. She said, "You all have this shared thing, but once one person in the family gets tested, then that person is on the island or in the water with the sharks, but they're not in the boat anymore. I've seen cases where someone got good news and called up the people they loved, and had a sibling hang up on them. Later, they apologized but said they immediately felt that as a result of the other person's getting a clean bill of health they were going to get it."[90] Sometimes people who test negative avoid siblings who have tested positive, thinking they would become resentful.

Genetic researchers themselves can wreak havoc with family relationships. One research group published a modified pedigree chart of a family in which they noted there were two children who had "paternal genotype inconsistencies."[91] This suggestion that two children might have been the result of an adulterous relationship had a disastrous effect on the immediate family.

In some families, siblings share a bond created by the risk of having a genetic mutation that runs in the family. When one sibling learns that he or she has the gene mutation and the other does not, they may feel less close. The impact on a sibling relationship can be even more profound when one sibling is actually affected with the disorder for which another is undergoing prenatal diagnosis. When the woman's brother or sister has a recessive disorder such as cystic fibrosis, for example, she may not want to learn her own genetic status because she believes that her risk for having the genetic mutation makes her seem closer to her affected sibling than if she learns she does not have the mutation.[92] By contrast, learning that her fetus is affected with a recessive disorder may strain her relationship with her affected sibling, who may view the woman's consideration of terminating the pregnancy as a rejection of the sibling.[93]

The Mattingly family has been affected by an inherited neuromuscular disorder, seemingly unique to at least 70 of 150 known descendants of Thomas Mattingly, who died in 1664. Organized by Andrew Mattingly, researchers hoping to locate the responsible gene, to eventually find a treatment or cure, have recently narrowed the search to the tip of the ninth chromosome.[94] One of the Mattinglys felt that, had a test for the gene been available, she would have taken it. She said, "I would not have had children. I don't think it would be fair. It's bad enough now, the way the whole family is, without passing it on further." Another Mattingly felt differently. Although one of her children had the disorder, she decided to have another child, despite the risk involved. She said, "It comes back to the basic premise that, just because everybody doesn't walk the same, their life is still valuable. I wish I didn't have [the disorder]. I wish my son didn't have it. It bothers me a lot. I'm starting to feel my world is limited by it, and that really bothers me, bugs me, irritates me. But I'm still glad I'm here."[95]

Genetic services can bring families together or wrench them apart. Old grievances and disputes surface in new ways around genetic issues. And new questions of responsibility need to be addressed by those individuals who are bound by blood ties.

Impact of Genetic Testing on Relationships with Spouses and Potential Spouses

For centuries, people have decided upon mates based on their potential for producing worthy children. Genetic testing can directly or indirectly affect that choice. It can also have a profound impact on intimate relationships.

When scientists at the turn of the century began to label certain traits as being inheritable — including pauperism and feeblemindedness — people began to make private choices about whom to marry on genetic grounds, in order to avoid having a child with a disfavored trait.[96] In 1910 the Eugenics Records Office, which trained field workers to collect family histories from people around the country, was established in Cold Spring Harbor, New York.[97] By 1924, data from about 750,000 people had been entered, and people made inquiries to the office about whether particular proposed marriages would be eugenically appropriate.[98] Also in the early eugenics movement, states adopted laws to prevent certain people from marrying each other (such as a "normal" individual and a feebleminded individual, or a black individual and a white individual) on the grounds that their children would be "defective."

Even today, decisions are made about procreation based on genetics. If an infertile couple seeks to create a pregnancy using an egg or sperm provided by a third party, medical professional guidelines require that the donor should be screened for certain genetic disorders so that he or she will not pass them on to the child.[99] One state — Ohio — also has a statute requiring sperm donors to be screened genetically.[100]

Currently, in the Orthodox Jewish community of New York, where arranged marriages are still common, genetic information is increasingly being taken into consideration at the matchmaking stage. People who are of Ashkenazi Jewish descent have a one-in-25 chance of having a Tay-Sachs genetic mutation; if two such carriers marry, each child has a one-in-four chance of having the devastating disease. A child with Tay-Sachs appears normal at birth, but later loses motor functions, suffers massive neurological deterioration and seizures, and generally dies by age three. A program in New York known as Chevra Dor Yeshorim (Association of an Upright Generation) offers Tay-Sachs carrier screening to Orthodox Jewish adolescents. Before a marriage is arranged, the matchmaker calls the program with the identification numbers of the two individuals. If they both carry the gene for Tay-Sachs, they are not matched for marriage. One rabbi has proclaimed, "It is the obligation of every parent, without exception, to turn to Chevra Dor Yeshorim, and heed their advice, before finalizing a match for his or her child."[101]

The Amish, too, are trying to find ways to prevent marriages that could result in the birth of a child with a serious genetic disorder. Genealogical studies have traced six genetic disorders prevalent in different Amish communities back hundreds of years to specific individuals, many of whom are ancestors of most of the present-day Amish. The Amish tradition of secluding their communities from the outside world and marrying strictly within their faith has genetically translated into a higher prevalence of genetic disorders in their children.[102] In fact, one study showed that Amish inbreeding accounts for 40 percent of the childhood deaths in the Amish community.[103] The Amish consider genetic diseases to be tragic, but they consider the birth of a child with a genetic disease to be a gift from God, and a natural part of life. Religiously, the Amish reject birth control and abortion, which might decrease the prevalence of children born with genetic disorders in their communities. But they have changed their rules regarding marriage in one major way in the face of the prevalence of genetically diseased children: first cousins are now forbidden to marry. Additionally, in an attempt to decrease the risk of genetic diseases in their grandchildren, some Amish are challenging tradition by arranging marriages to Amish living in other communities who have descended from other families.[104]

Most of us do not consciously seek a partner based on his or her genetic pedigree. Yet increasingly such information is becoming available and creating complicated interpersonal questions. If a person has a gene for, say, colon cancer, should he or she warn a potential spouse? What happens when the recipient who initially believed such information would not make a difference in the relationship finds that he or she is beginning to view the loved one differently?

When each partner has a gene for a recessive disorder, there is a sharing of any guilt and feeling of responsibility. But this is not true of all genetic disorders. X-linked ones, such as Fragile X, and dominant ones, like Huntington's disease, are passed on by only one parent. There is already evidence from testing programs about the guilt and disruption this imbalance can cause. Women who pass on Fragile X, causing mental retardation in their sons, feel an enormous guilt about failing in what they see is women's traditional role—giving birth to healthy males.[105] And some people who learn they have a gene for a dominant disorder such as Huntington's disease tell their spouse, "You can divorce me."[106]

During a subsequent pregnancy, Maureen Schmidgall learned her first child had Fragile X. She says, "It's hard not to blame yourself." Once she and her husband found out that their first child had the illness, they went through the rest of the pregnancy in fear. They decided to act on that information—and will not have any more children.[107]

MENTAL HEALTH, GENETICS, AND THE DECISION TO REPRODUCE

Although Clea Simon, 35, is healthy, her brother and sister both had terrible cases of schizophrenia that caused them to be delusional and psychotic. Clea's brother jumped off a cliff and died when he was 30, and her sister used to threaten Clea's life. Clea's parents tried to keep the illness from their neighbors to protect Clea from any social or psychological problems that may have resulted, but those who knew treated Clea as if she were "damaged." And even though her parents shielded her, she is still haunted by the illness of her siblings.

Several years ago, Clea was dating a man for a few months until he turned to her and said, "The problem is I'm looking for someone I could get seriously involved with, and this schizophrenia thing really scares me. You see, I want to be able to have children, and you're too much of a genetic risk." After her astonishment at her date's sentiments wore off, Clea felt scared by the thought of jeopardizing the health of her future children.

Clea researched schizophrenia and found out that her children would be eight times more likely to develop it than the average person, although the chances were still a low 3 to 8 percent. Clea says, "I can too easily envision a mixture of guilt and grief. Because of what I know about schizophrenia, my mothering could be polluted by anxiety." Clea wonders what kind of effect that would have on her child. Clea describes, "Sometimes, I am stopped cold by the realization that I might end up being afraid of my own child—that she or he could grow up to be like my siblings." She explained, "In this way, the fear creeps back in. That because of my past, I am dangerous, or at least not worthy of happiness. These insecurities are prompted by guilt, of course. I am the only one of us three who even has the luxury of making a decision about whether to have a child."

SOURCE: Clea Simon, "Haunted by My Family's Madness," *Washington Post*, March 9, 1997, at C01.

Dan and Nina Butler have had five children, three of whom have genetic disorders. The oldest one has cystic fibrosis, another Fragile X, and the third a developmental delay similar to autism. The Butlers admit they probably would not have gotten married had they known that they were both carriers of cystic fibrosis. "We would be worse off for that," they say. They assert that if prenatal testing could have informed them of their children's disabilities, they would not have taken the tests. The Butlers say, "We have absolutely no regrets about our choices."[108]

A survey of 214 women attending OB/GYN clinics in Ohio found that 9 percent of respondents felt that cystic fibrosis carrier status was a good reason for a couple to avoid marriage.[109] Men are more likely than women to say they would alter marriage plans if they learned that their fiancé was the carrier of a recessive genetic disorder. Eight years after having participated in Tay-Sachs testing, 95 percent of female carriers responded that they would not alter marriage plans upon discovering that their partner or intended partner was also a carrier. In contrast, only 69 percent of male carriers responded definitively that they would not alter marriage plans if their intended spouse was also a carrier.[110] Another study of Tay-Sachs testing found that 25 percent of carriers and 6 percent of carriers' spouses felt that knowing their own or their spouse's carrier status would have affected their marriage decision.[111]

What people say they *will* do is not always what they actually do. A study of sickle cell anemia testing in Orchemenos, Greece, provides information on actual behavior.[112] People who are carriers of the gene are healthy themselves, but if they procreate with another carrier, every child has a 25 percent chance of having sickle cell anemia. The health care providers thought that the testing they offered would decrease the number of affected children by causing people to make more "rational" reproductive decisions (since, if a carrier and a noncarrier have a child together, there is no chance that the child will be affected with the disorder). What actually happened, though, was that carriers were stigmatized. The birth rate of affected children did not decrease since in some instances, the only person who would marry a carrier was another carrier.

In the Orchemenos study, 20 percent of normal parents and 10 percent of the carriers' parents advised their children to avoid marrying a carrier.[113] Seven percent of noncarriers avoided marrying a carrier and some broke off their engagement when they learned the potential spouse had sickle cell trait. Of the sickle cell carriers, 25 percent concealed their status from their potential spouse. One in five broke their engagement when they learned their prospective spouse was also a carrier.[114]

Sexuality, too, may be affected by genetic knowledge. A study of daughters of breast cancer patients found that satisfaction with sexual relationships was adversely affected by fears or worries about breast cancer. When compared to matched controls, daughters of breast cancer patients reported less frequency of sexual activity and less satisfaction with sexual encounters.[115] A similar reaction might occur in women who learn of their increased risk through genetic testing of themselves or their fetuses rather than from their mothers' illnesses. Another study assessing changes in relationships after predictive testing for Huntington's disease found that individuals who received an increased-risk result indicated a significant decline in their satisfaction with their primary relationship during the two-year follow-up period after receiving the test results.[116] Follow-up on twenty-one couples who had gone through Huntington's disease testing found that six had divorced, with three specifically attributing it to the testing.[117]

Just as individuals bring certain expectations to genetic testing, so, too, do couples. Genetic services can affect the emotional well-being of the partner of the person using those services. In one couple, when the wife was tested for Huntington's disease and learned that she did not have the mutation, she was disappointed. She thought that if she had the gene and would die anyway, she would have found the courage to leave her husband.[118] In another couple, the husband developed clinical depression when his wife learned she did not have the Huntington's disease mutation. He had arranged his life—and his retirement—so that he could care for her when she fell ill. He had been a workaholic before he retired and planned a "second career" as his wife's caretaker. When he found out that she did not have the mutation, he felt as if he had lost his job.[119]

Andrea Zankow, a genetic counselor at the University of California in San Francisco who studied forty-nine families with members who were tested for the Huntington's disease gene, reported: "We found a high portion of children that came from disorganized homes. These are kids growing up in a home with a parent who becomes increasingly inappropriate and ineffectual. The other parent often was overwhelmed, or absent, or alcoholic, and so the kids were really raising themselves, and they didn't have the role models for daily skills or coping mechanisms."[120]

If a person is found to have a genetic mutation associated with a serious genetic disease, his or her partner will naturally worry about the children they have created. Partners of individuals with Huntington's disease may develop resentment and hostility over the fact that the disease may have been transmitted to their children.[121]

Genetic information can also radically change a spouse's sense of himself or herself. Consider a man who learns that his wife has the genetic condition of testicular feminization. That means the wife has the chromosomes of a male, but her "male" reproductive organs have developed internally, rather than externally.[122] She has a vagina and appears to be a woman, but since she does not have internal female reproductive organs, she is infertile. There are a variety of ways in which the man might receive this news. He might be told the information by his wife after she has been tested. The wife's doctor might breach confidentiality and disclose the information to the husband against the wife's wishes; the doctor may believe that the information is "relevant" to the man since it means he will not be able to have a child with this wife. The man might learn the information after the wife's death if he takes advantage of any of the numerous offers to save a loved one's DNA and have it tested. Whether it should or not, the husband's sense of identity may be damaged by the news that he has been living with someone who is genetically a male.

When a person encounters a problem in life, he or she often turns to a partner for comfort. In the case of genetic testing, however, the partner himself or herself might be having a troubling reaction, different from that which the tested individual is experiencing. While married individuals typically are better adjusted than unmarried people, research on Huntington's disease testing found that married positives were less well adjusted than unmarried positives: "Unlike single persons, married positives have the added psychological stress of knowing that they are 'causing' distress in someone close to them," notes a group of researchers at Johns Hopkins. "Alternatively, positives may have been affected by their spouses' reaction to the news."[123]

While working at Johns Hopkins Hospital in Baltimore, Kimberly Quaid counseled people for months before they took the genetic test that would reveal whether they had the gene for Huntington's disease. She often worked with people who were married, and she found that unfortunately "a lot of people don't get the family support that they expect" when they receive this kind of predictive genetic information. Quaid stated: "One husband told me he probably wouldn't be able to handle it if his wife tested positive. He might leave her and would have to live with the knowledge that he was a coward. When I asked another man what he would do if his test were positive, he said: 'Well, I wouldn't cheat on my wife.'"[124]

The use of genetic services can cause people to lose their internal moorings, to view themselves in a different way. It can also make conditional some of the previous unconditionals in relationships, creating fault lines in the very foundation of people's lives.

4   The Changing Face of Parenthood in the Genetics Era

Genetic services, like most medical services, affect people profoundly as individuals. But they also affect people as members of couples and as parents of children. The central tenet of genetics—inheritability—adds even more difficult and complicated issues as people make decisions not only for themselves but for future generations.

Impact on Reproduction

Family history has long been used to predict a couple's general chance of giving birth to a child with a particular disorder. Starting thirty years ago with the advent of prenatal diagnosis through amniocentesis, specific genetic information about fetuses became available to parents. Today genetic information related to reproduction is obtainable in numerous ways. One partner may have a mutation associated with Huntington's disease or breast cancer, for example, and face a decision about whether the fetus should be tested for that mutation. In other instances, the parents may not know their genetic makeup, but each (or maybe both) may be tested in advance of reproduction (or even during pregnancy) for mutations that are common in their ethnic group—such as cystic fibrosis for Caucasians, sickle cell anemia for African Americans, or Tay-Sachs for Ashkenazi Jews. Or the fetus itself may be tested, which can reveal previously unknown genetic information about the mother, the father, or both. Parents-to-be may learn that both are carriers of a recessive disorder, that one has a mutation associated with a dominant disorder, or that the woman has passed on a mutation associated with an X-linked disorder.

The range of conditions that can be screened for prenatally is growing exponentially each year. More than five hundred different conditions can be diagnosed through chorionic villi sampling or amniocentesis.[1] The availability of these tests affects even those people who decide not to have them done. Some women say that friends and relatives have made them feel irresponsible for not having genetic testing; others have felt guilty and responsible if they had a child with a genetic disorder (either after refusing testing or after deciding to carry through the pregnancy of an affected child). Yet, as more and more prenatal genetic tests become available, parents are increasingly feeling that they are put into the position of playing God. Should they have prenatal screening for a disorder that won't affect their child until much later in life—or should they bank on a cure being developed in the child's lifetime? Should they abort a fetus whose disorder is treatable, albeit at some expense? As testing becomes available to tell whether a fetus is at a higher likelihood of suffering from breast cancer, colon cancer, heart disease, diabetes, and Alzheimer's disease, should such tests be utilized? What about testing for alcoholism, violence, and other behavioral traits? In studies with varying degrees of scientific repute, genes have been implicated in shyness, bedwetting, attempted rape, homosexuality, manic-depressive disorder, arson, tendency to tease, traditionalism, tendency to giggle or to use hurtful words, and zest for life. Should testing for such traits be done prenatally? If so, what should be done with the resulting information?

The decision of whether or not to terminate a pregnancy is a personal choice that is influenced by many factors. With increasing frequency, women or couples are faced with making a decision about which genetic defects are severe enough to warrant the termination of a pregnancy.[2] In a Japanese study questioning parents of young cancer patients, 27 percent said they would undergo prenatal genetic testing for cancer, 28 percent would not, and 38 percent were undecided.[3] A study by the Harris organization found that 89 percent of Americans support prenatal testing for severe or fatal genetic diseases, but that there is little support for genetic testing for minor or "cosmetic" reasons.[4] The question is, What is a severe genetic disease?

At least three factors may influence the parents' perception of severity. The first is the impact the condition has on the quality of the child's health, including survival, suffering, and limitations in function and activity. Tay-Sachs disease, for example, is severely debilitating in infancy, and death occurs at a young age. The second factor is the age of onset of the disease. Some diseases appear at birth, and others take years to manifest—diseases in the second group are often considered to be less severe. The third factor is the "probability that the genotype will influence the phenotype"—the extent to which the genes influence the health of the child. Turner syndrome causes the individual to be short and infertile. In this situation, different parents may have different views on whether short stature and infertility are considered serious health problems.[5]

Many couples who use genetic services in conjunction with reproduction feel that it has offered them an overall benefit by allowing them to make an informed choice about their pregnancies. In fact, couples have sued when they felt deprived of prenatal genetic information. At least one federal court has recognized the importance of this choice by indicating that the constitutional protections of the abortion decision logically "must also include the right to submit to a procedure designed to give information about that fetus which can then lead to a decision to abort."[6] In most states, couples can bring medical malpractice suits known as "wrongful birth" suits if they give birth to a child with a serious genetic disorder and their physician failed to advise them of the availability of prenatal genetic testing that was appropriate to individuals of their age, ethnic background, and family history.[7] Wrongful birth suits can also be brought when a mistake is made in prenatal testing or the interpretation of results and the couple is erroneously told that their fetus does not have a particular genetic anomaly.[8]

The fact that many couples undergo prenatal testing and make decisions based on that information, however, does not mean the process is an easy one.[9] The testing generally provokes anxiety, whether or not the couple ultimately receives the comforting news that the fetus is not affected with the disorders for which it has been tested. The news that the fetus is affected with a disorder has even more profound consequences as the couple grapples with the issue of whether or not to continue the pregnancy. Moreover, abortion of the wanted fetus because of its genetic profile often causes grief, sorrow, and long-term guilt. And the diagnosis that the fetus has a particular genetic disorder often has an impact that lasts far longer than the pregnancy, as it provides the couple with information about their own genetic statuses. The couple will face varying degrees of difficulty incorporating this information about themselves into their self-concepts, personal relationships, decisions about whether to have other children, and dealings with social in-stitutions.

When amniocentesis first came into use thirty years ago, abortion of an affected fetus was viewed as a temporary approach that would soon give way to treating affected fetuses.[10] Not only have such treatments not materialized, they are not being aggressively pursued. "In fact," notes McGill University epidemiology professor Abby Lippman, "in the current sociopolitical climate of North America, where individual responsibility to prevent health problems takes precedence over social responsibility to support policies that promote the general well-being of all, developing remedies is probably far less likely than developing ways to prevent the birth of those who may have such problems."[11] The developments in diagnosis have become more elaborate, pushing testing back to earlier in the pregnancy, through techniques such as chorionic villi sampling, or even pre-implantation embryo screening.

The Physical Risk of Prenatal Screening

Some forms of prenatal testing present physical risks to the fetus, but others are relatively unintrusive. Sampling blood from the fetus while it is in utero through fetoscopy, one of the first means of prenatal tests developed for fetuses, is associated with a 3 to 6 percent risk of fetal death.[12] Amnio-centesis, in which fluid from the amniotic sac is withdrawn and analyzed, causes spontaneous abortions in approximately one to two in every 1,000 pregnancies.[13] Chorionic villi sampling (CVS), in which tissue surrounding the fetus is sampled and analyzed between eight and twelve weeks gestation, is associated with a spontaneous abortion rate of about 3.6 percent.[14] In addition, CVS in early pregnancy presents a risk of limb deformities in approximately one in 3,000 cases.[15] These procedures entail physical risks to pregnant women as well, particularly risks of infection.[16] Some women have chosen to undergo prenatal screening despite the risks because they intend to terminate a pregnancy if the fetus is diagnosed as having a serious disorder.

The prenatal diagnostic techniques of fetoscopy, amniocentesis, and cho-rionic villi sampling have generally been limited to women whose fetuses are at greater-than-average risk of having a genetic or chromosomal disorder, such as women over age thirty-five (who are at a greater-than-average risk of giving birth to a child with Down syndrome) or women with family histories of or ethnic-group familiarity with genetic disorders. In many instances the women being offered such prenatal testing have had some experience with the disorders for which the fetus is being tested. Most people are familiar with Down syndrome. Those who have a family history of a disorder such as cystic fibrosis or sickle cell anemia have an affected relative and thus are likely to have some knowledge about the disorder. And in ethnic groups that have a higher-than-average risk of certain genetic diseases, there is often widespread knowledge among members of the community about the particular disorder.

Now, however, prenatal testing is being developed that is less physically risky to the fetus and to the pregnant woman. Some testing can be performed with a sample of the pregnant woman's blood. Genetic mutations affecting the fetus can be identified through maternal/fetal cell sorting, a technique that can analyze fetal cells circulating in the woman's blood. Since the physical risk has been reduced, women who might not otherwise have sought prenatal testing are being asked to do so. One result is that women who have no personal knowledge or experience with a disease or disorder are faced with having to decide whether they are willing to raise a child with these conditions. This stands in contrast to those situations in which such disorders are generally understood within the woman's community. The more accessible prenatal testing becomes, the more important it is to adequately inform and educate potential parents about their fetus's condition as well as about the long-term consequences of whatever decisions they make.

Impact on Pregnancy

The use of prenatal screening changes pregnant women's relationships with their fetuses. The existence of prenatal testing may be turning women's perception of pregnancy from that of a normal, healthy experience into a pathological condition. Even though only a small fraction of women learn that they are carrying a fetus with a genetic abnormality, women now think of their pregnancies as being at risk. In one study, women overestimated the chances of an abnormal pregnancy, and that concern was not diminished through genetic counseling and education.[17] "The concept of risk dominates the process of becoming a mother in North America today," notes Lippman. "By attaching a risk label to pregnancy, physicians reconstruct a normal experience, making it one that requires their supervision."[18]

Instead of bonding with their fetus early in the pregnancy, women who undergo prenatal testing delay bonding until they learn the results of the testing. They often do not tell friends or family members about the pregnancy until they receive the genetic testing results. Overall, this results in what Baruch College sociologist Barbara Katz Rothman has called "the tentative pregnancy."[19]

One major purported benefit of prenatal testing is "reassurance." But this may not prove to be the case. Of women who received normal results from amniocentesis, 20 percent remained concerned about the possibility of fetal abnormalities. Of women who declined testing, 20 percent were similarly worried.[20] Thus, prenatal testing does not necessarily lead to a net gain in reassurance.

In fact, the testing itself produces anxiety. Consider one example: Approximately 4,000,000 babies are born in the United States each year. Since neural tube defects have an incidence of one in 2,000,[21] there are potentially 2,000 affected infants per year.[22] A blood test is offered to most pregnant women to screen their fetuses for this problem. Originally that blood test, the maternal serum alphafetoprotein test for neural tube defects, had a high false-positive rate, with 5 percent of women (200,000) appearing to be positive. That means that 198,000 women unnecessarily went through the stress of worrying about their fetuses and having follow-up testing. Moreover, the 2,000 women who were true positives had no indication of how serious their child's health problems would be.

A more intrusive prenatal test, amniocentesis, has become so routinized, with so many women receiving negative results, that the couple who learns the fetus has a genetic mutation may feel stigmatized. "I felt ashamed and embarrassed, somehow freakish, that this had happened," said one woman. "I felt that others would be shocked to learn that we had produced an abnormal baby, that we would be outcasts."[23]

Often women are concerned about how their mates and others will judge the actions they take during pregnancy. The woman may believe that the father of the fetus will blame her if something goes wrong with the pregnancy.[24] Some women who would continue a pregnancy of an affected fetus, particularly a fetus with a milder disorder such as borderline retardation or anomalies that prevent the child from maturing sexually or becoming fertile, abort because they do not think their husband would want such a child.[25] Dorothy Wertz observes: "Men find it more difficult than women to accept a boy who will not turn out to be a 'man' according to accepted social definitions."[26]

Impact on Abortion

• Women who learn that their fetus has a genetic mutation may feel compelled to abort (or even pressured by their physicians to do so), particularly since, as Wellesley professor Adrienne Asch points out, society portrays people with disabilities "as being permanently ill and in pain."[27] Some women do not feel that they have a real option to continue a pregnancy of a child with Down syndrome, since society does not accept and nurture children with disabilities.[28]

A survey found that 39 percent of people felt that every pregnant woman should have prenatal testing and 22 percent felt that a woman should be required to abort if the baby has a serious genetic defect.[29] At the same time, women may feel that they will be judged too harshly by friends and relatives for aborting. Women who terminate an affected pregnancy after prenatal diagnosis often feel guilty.[30] "I have had to face the fact that we *chose* to kill this fetus, which we had conceived in love and hope," said one woman who aborted after prenatal diagnosis.[31] A woman's perception of herself may change after such an abortion, causing her to feel like an "agent of quality control in the reproduction production line."[32]

Couples are very careful in deciding whom to tell all the facts surrounding the termination of the pregnancy, since they are concerned about the "value judgments" made by others who have never gone through the same experience.[33] "The other side of the coin is that when people don't say anything, or assume that it's over and that we're 'all better now' or tell us how 'lucky' we are—that's like denying our anguish," said a woman who aborted after prenatal diagnosis. "Now this pregnancy has been wrenched and torn out . . . and it *hurts*. I'm a big open wound, dripping torn tissue and blood all over. I wonder sometimes that other people can look at me and talk to me, ignoring the blood."[34]

Women who have aborted affected fetuses express anxiety over whether or not they should attempt to conceive again. The majority indicate that they fear having to go through the whole experience again. In fact, some women indicate that even if, in future pregnancies, the prenatal testing results indicate the fetus does not have the tested-for disorder, they will still be very anxious about the occurrence of defects that are not detectable by testing.[35]

Women who terminate pregnancies in the second trimester after prenatal diagnosis suffer psychological traumas similar to women who miscarry in the second trimester.[36] In some instances their grief may be even more pro-nounced, since the ultrasound image that accompanies prenatal diagnosis may create an even greater emotional bond to the fetus. Some women who have a pregnancy terminated because of a serious genetic disorder "decline prenatal testing altogether in later pregnancies because of their feelings of remorse," according to geneticist Angus Clarke. He notes: "The lives of some of these families might have been disturbed less by the birth of an affected child than by the termination of pregnancy."[37]

Sociologist Barbara Katz Rothman points out that there is no social sup-port for the grieving woman who terminates a pregnancy for genetic reasons, while there are lots of social supports for a woman whose child is stillborn. She observes that some women who abort fetuses with anencephaly might have been less harmed psychologically if they had waited three months and gone through the stillbirth. Then, traditional social supports for comforting grieving mothers would have come into play. She notes that these women may not have been benefited as much by the genetic testing as geneticists think.

Impact on the Decision to Carry the Pregnancy to Term

Women may also feel guilty when they carry an affected pregnancy to term. This is particularly true in the case of mothers who already have chil-dren with a genetic disorder and who give birth to additional children with the same disorder.[38] The offer of genetic testing itself implies a tacit rec-ommendation to abort.[39]

The new genetics is thought to offer new choices, but at times the mere existence of a technology contains an implicit coercion to use it. "Offering carrier screening to assist couples in making reproductive decisions is not a neutral activity but, rather, implies that some action should be taken on the basis of the results of the test," write Sherman Elias and George Annas. "Thus, for example, merely offering screening for a breast-cancer or colon-cancer gene suggests to couples that artificial insemination, adoption, and abortion are all reasonable choices if they are found to be carriers of such a gene."[40]

Sometimes the coercion is more than implicit. At some facilities in En-gland, the compensation of genetic counselors is based on the number of women who undergo testing. In addition, for the woman who does undergo prenatal testing, there is "the possibility that her medical advisors might wish her to terminate the pregnancy if this would improve their audit returns" or demonstrate that the cost of testing has been "well spent" by "preventing" the birth of children with disabilities.[41]

Society may make women feel guilty for continuing the pregnancy of a fetus with even a slight disability. This occurred in the case of Bree Walker-Lampley, a California anchorwoman affected with ectrodactyly, a mild genetic condition that fused the bones in her hand. When she decided to continue a pregnancy of a fetus with the same condition, a radio talk show host and her audience attacked the decision as irresponsible and immoral.[42] Lampley, along with several disability-rights groups, filed a Federal Communications Commission complaint against the radio station for violating the federal personal attack rule and failing to present both sides of the issue.[43] The complaint was denied.

Impact on Future Use of Genetic Services

The fear of undergoing involuntary genetic testing may deter women from seeking medical care during pregnancy, which may cause physical risks to them and their fetuses.[44] In addition, a mother's stress from being tested against her will and learning information she does not want to know may actually harm the fetus in utero.[45]

Additionally, the psychological impact of receiving genetic information through testing may be sufficiently troubling that it will lead some people to refuse future testing in situations in which the testing might be beneficial. In one cystic fibrosis carrier screening program, women were sufficiently troubled by the process that they refused subsequent prenatal testing for maternal serum alphafetoprotein.[46] In addition, some couples who terminated their pregnancies as a result of Huntington's disease testing on the fetus, refused prenatal testing altogether during their subsequent pregnancies and carried the fetuses to term.[47]

Effect on Relationships with Other Children

Unique issues are raised when a couple already has a child affected with a certain disorder, such as cystic fibrosis or beta thalassemia, and learn that their fetus is also affected. For some, abortion is a symbolic rejection of the existing child. Some women actually ask permission of the first child to abort the affected fetus. "It's as if they want to share their guilt with another person," says one psychologist. "Usually, this only occurs if the existing child is 15 or older, but there have been instances in which it has occurred with a child as young as age 9."

Effect on Relationships with the Resulting Child

The information that couples receive about their fetus can affect their expectations about that child. Stereotyping about the child now begins even before the baby is born. Often, as a by-product of genetic testing, couples learn the sex of the fetus. Mothers who learn that they are carrying a female fetus characterize the fetus in utero as making gentle, slow movements that are lively but not excessively energetic. In contrast, mothers of sons describe their movements as very strong, vigorous, a series of earthquakes, not altogether pleasant.[48]

Parents may learn genetic information about their fetus that does not lead them to abort, but may affect how they treat the child as they raise him or her. On the one hand, parents may overindulge or overprotect a child with a gene for a late-onset disorder such as colon cancer or Huntington's disease. On the other hand, parents may devote fewer resources to education of that child, on the grounds that he or she may not have a long professional career.

The birth of a child with a disability used to be viewed as a normal risk of reproduction, or even as God's will. Now, increasingly, it is viewed as a medical fault of doctors who did not offer genetic testing or parents who refused testing or proceeded with the pregnancy. In California, New Jersey, and Washington, a child born with serious genetic disorders can sue a physician or genetic testing laboratory for wrongful life if an error was made in prenatal genetic testing and his or her parents were erroneously told that he or she did not have a genetic anomaly.[49]

A California case, *Curlender v. Bio-Science Laboratories*,[50] took the concept of wrongful life one step further, with the startling suggestion that children could sue their parents. The court indicated that a child born with Tay-Sachs, a genetic disease, could bring suit against her parents for not undergoing prenatal screening and aborting her. The California legislature immediately passed a law prohibiting suits against parents, but society at large—from friends and relatives to health insurers and physicians—is increasingly pressuring parents to take genetics into account in making childbearing decisions.

Impact on Decisions About Children

Genetic testing has become routine during pregnancy. Some parents, though, have gone further and sought genetic testing of existing children. A few years ago, a mother entered a Huntington's disease testing facility with her two young sons. "I'd like you to test my sons for the HD gene," she said. "I only have enough money to send one to college."[51] That request and similar requests to test young girls for the breast cancer gene or other young children for carrier status for recessive genetic disorders raise enormous questions about whether parents should be able to learn genetic information about their children and whether it will cause them to treat the children differently. Such a request is far different from prenatal screening where the outcome is abortion. Here, the child who has already been born may be treated differently based on genetic information.

Genetic testing of children is fundamentally different from genetic testing of adults. Currently, adults who participate in genetic testing, whether predictive or carrier testing, do so as a matter of choice. When children undergo genetic testing, typically it is the parent who chooses to authorize the genetic testing on the child.

Some parents clearly want access to genetic testing. A majority of the respondents in one survey felt that testing children for late-onset disorders was acceptable. In addition, 61 percent thought parents should be allowed to have their children participate in predispositional testing for Alzheimer's disease. Eighty-four percent would have a thirteen-year-old daughter tested for her predisposition to breast cancer. Moreover, 47 percent of those who thought parents should be allowed to test children indicated that they would disclose predispositional test results to their children.[52]

Dorothy Wertz and Philip Reilly surveyed members of Helix, the principal U.S. network of genetic testing laboratories. Twenty-three percent had tested children under age twelve for Huntington's disease. More than 40 percent had tested based on a direct parent request, rather than a request from a physician.[53]

The aggressive marketing of genetic tests pressures physicians—and parents—to consider testing children. The manufacturer of genetic tests for melanoma predisposition sent letters to dermatologists urging that "early screening with this easy and painless test is particularly useful when testing children."[54]

The existence of genetic testing may sometimes help parents identify and treat serious disorders in their children. The classic case is phenylketonuria, in which an affected infant will be profoundly retarded unless he or she is put on a low phenylalanine diet shortly after birth. Other uses of genetic testing in children are less clearly therapeutic, however. Parents may seek to have asymptomatic children tested for late-onset disorders for which there are no effective preventive or treatment strategies that must be started at an early age. Young teenagers, too, may pressure doctors for genetic testing, especially after seeing the effects of a severe disease (such as breast cancer) on their relatives.[55] Yet there is little understanding of the impact on the parents—and on the child—of providing such information.

Impact on Parents' Emotional Investment in Their Children

There are major psychological implications of allowing parents to learn their child's genetic makeup. "A child known to have a deleterious gene may be overindulged, rejected or treated as a scapegoat," according to a statement on genetic testing of children prepared by the American Society of Human Genetics (ASHG) and the American College of Medical Genetics (ACMG).[56] Giving parents presymptomatic genetic information may cause them to overinvest or underinvest emotionally and financially in their children. The parents may be less likely to devote emotional care and financial resources to a currently healthy child who, in ten, twenty, or even fifty years, will fall ill.

Once test results are obtained, they follow the child throughout life. " 'Planning for the future,' perhaps the most frequently given reason for testing, may become 'restricting the future' (and also the present) by shifting family resources away from a child with a positive diagnosis," wrote Dorothy Wertz, Joanna Fanos, and Philip Reilly, an interdisciplinary team assessing the issue, in the *Journal of the American Medical Association*.[57] Such a child "can grow up in a world of limited horizons and may be psychologically harmed even if treatment is subsequently found for the disorder."[58] The joint ASHG/ACMG statement notes, "Presymptomatic diagnosis may preclude insurance coverage or may thwart long term goals such as advanced education or home ownership."[59]

The impact of genetic information is sufficiently potent that some parents are affected by the mere hint that their child has a genetic mutation, even if later tests demonstrate that this is not the case. That is exactly what happened in the public health newborn screening programs. Screening by its nature is overly broad; in newborn screening for cystic fibrosis, for example, "only 6.1 percent of infants with positive first tests were ultimately found to have cystic fibrosis on [confirmatory] sweat chloride testing."[60] Yet one-fifth of parents with false positives on newborn screening for cystic fibrosis "had lingering anxiety about their children's health."[61] In a pilot Wisconsin newborn screening program, 5 percent of the parents whose children had initial positive CF tests that were later disproven still believed a year later that their child might have cystic fibrosis.[62] Such a reaction may influence how parents relate to their child. University of Wisconsin pediatrician and bioethicist Norman Fost reports that in the Wisconsin newborn screening for cystic fibrosis, of the 104 families with false positives, 8 percent planned to change their reproductive plans and an additional 22 percent were not sure whether or not they would do so. In fact, in France, the newborn screening program for cystic fibrosis was terminated at the request of parents who objected to the high number of false positives.[63]

In addition to tests of existing children, parents obtain genetic information about their children when they subject the fetus to prenatal screening and then carry the pregnancy to term. Some clinics refuse to undertake Huntington's disease testing on children, but they provide such testing on fetuses with the thought that women will likely abort such fetuses. Since some women choose to carry the child to term rather than to abort,[64] those women will have information about their children that professional guidelines recommend not be given to parents. In addition, some women will learn that their fetuses will not be affected with a recessive disorder such as cystic fibrosis, Tay-Sachs disease, or sickle cell anemia, but instead will be carriers of the disorder. Even though carrier status will not affect the child's health, the parents may nonetheless view the child as at risk medically (because of this genetic "abnormality")[65] and may be overprotective of the child or overconcerned about the child's health.

Impact on the Child's Well-Being

Given that genetic information has been shown to have a profound effect on many adults who receive it, it is reasonable to assume that it will also have a powerful effect on children. A child who knows about the presence of a genetic mutation may feel "defective" or "different." The ramifications may turn out to be especially problematic if the person receives this information in adolescence, when trying to fit in with friends and feel confident about himself or herself are major concerns.

The complex factors that will affect children's emotional reaction to genetic testing are poorly understood. Researchers are only just beginning to study the matter. In one such study, forty-one children six to sixteen years old were tested for familial adenomatous polyposis (FAP), a treatable and possibly preventable form of hereditary colon cancer. Nineteen children tested positive for the mutation, and twenty-two tested negative. Three months after testing, all forty-one children who had mothers with FAP had significantly higher anxiety scores than children whose fathers had FAP. Mutation-positive children with affected mothers also had significantly higher scores for depression. Regardless of their test results, the groups with affected fathers had a significant decrease in anxiety scores at follow-up, and those with affected mothers had a significant increase. Although more research needs to be done in this area, these results suggest that when children undergo predictive genetic testing, their psychological reactions may be less favorable if their mother has the disease than if their father has the same disease.[66]

Some parents say they consult their children before subjecting them to genetic testing. One mother who has familial adenomatous polyposis drove her children—ages nineteen, twelve, and ten—from their home in North Carolina to Maryland for genetic testing for the disease. In the car, she said, "If you don't want to know, tell me now."[67] But is it conceivable that a child in that situation would have felt that he or she could have refused—particularly if other siblings were planning to take the test?

If genetic testing reveals that the child has a mutation that could result in illness later in life, the child may be subjected to intrusive, untested, or risky interventions. It has been suggested that children be tested for Li-Fraumeni syndrome, which creates a 30 percent risk of developing cancer by age thirty. The benefits of testing children for Li-Fraumeni syndrome have not been proven. The idea is that early detection will lead to better outcomes, but Diane Hoffman and Eric Wulfsberg criticize the Li-Fraumeni recommendations on testing children as being based on "unsupported optimism,"[68] since most Li-Fraumeni cancers are incurable and "no evidence indicates that knowing a child carries the mutant gene is of any benefit."[69] Norman Fost notes that, in the case of Li-Fraumeni, "these theoretical benefits must be weighed against the risk of the possibly incapacitating psychological trauma associated with growing up under the sword of Damocles."[70]

Discovery that the child has a gene that predisposes him or her to colon cancer may lead to many physically intrusive colon exams. The physical and psychological impacts of such surveillance mechanisms in children have not been studied. Certain interventions may create more risk than benefit; for example, the use of monitoring X-rays for individuals with a breast cancer mutation may actually cause cancer in some individuals who would not otherwise have gotten it.[71]

Testing children is appropriate in limited circumstances—if the test will detect conditions that are readily preventable or treatable. "Timely medical benefits to the child should be the primary justification for genetic testing in children and adolescents," according to an American Society of Human Genetics/American College of Medical Genetics joint statement.[72] Health care providers sometimes overestimate the benefits of therapy, however. Often, early detection has no medical value and can harm the child and other family members by causing significant psychological stress before the time the child becomes symptomatic.[73]

Concerns also arise because we do not know if proposed preventive strategies will work. The National Advisory Council for Human Genome Research recommended *against* population screening for breast cancer in part because the effectiveness of preventive measures has not yet been documented.[74]

Impact on the Child's Behavior and Limitations on His or Her Adult Life

The possibility that genetic testing of children may lead to a dangerous self-fulfilling prophecy led to the demise of one study involving testing children. Harvard researchers proposed to test children to see if they had the XYY chromosomal complement, which had been linked (by flimsy evidence) to criminality. They proposed to study the children for decades to see if the ones with that genetic makeup were more likely to engage in crime than those without it. They intended to tell the mothers which children had XYY. Imagine the effect of that information—on the mother, and on the child. Each time the child took his little brother's toy or lashed out in anger at a playmate, the mother might freeze in horror, wondering if her child's genetic predisposition was revealing itself. She might intervene when other mothers would normally not, and thus distort the rearing of her child.

Presymptomatic testing of children will have an impact on the child's adult life, even in the years before the disease manifests. The child may be subjected to pressures from parents about major life choices and may be discriminated against in education, employment, and insurance eligibility. The ASHG/ACMG statement notes: "Expectations of others for education, social relationships and/or employment may be significantly altered when a child is found to carry a gene associated with a late-onset disease or susceptibility. Such individuals may not be encouraged to reach their full potential, or they may have difficulty obtaining education or employment if their risk for early death or disability is revealed."[75]

Parents may unwarrantedly interfere with the child's later choice of a mate or decision to reproduce. One study that examined decision making in families with cystic fibrosis found that 31 percent of mothers and 32 percent of fathers who had children with cystic fibrosis (each of whose *other* children have a 50 percent risk of being a carrier) believed carrier testing was important to provide information pertaining to risk that would aid them in their decision making concerning marriage and reproduction. Twelve percent of mothers and 18 percent of fathers thought carrier testing was important to avoid marriage of carrier couples and/or avoid children in couples where both partners were carriers. Fifteen percent of mothers and 14 percent of fathers indicated that testing was important to avoid the birth of individuals with cystic fibrosis.[76]

A major concern of those parents who choose genetic testing for their children is that the children have knowledge of their carrier status before they become sexually active and that they be fully informed of their genetic risk before they marry.[77] In studies of parents of individuals with Fragile X syndrome, reproduction and relationships emerge as the areas of greatest concern.[78] Seventy-two percent of the parents were very concerned that their children know the genetic risk before becoming sexually active, and 63 percent of the parents were also very concerned that their children be able to marry, fully informed of their carrier status.[79]

Impact on Other Family Members

Genetic information about a particular child can have a negative psychological impact on other family members. Sweden began a national public health program screening newborns for alpha 1-antitrypsin deficiency[80] with the idea that if parents were aware of the problem they could keep their children away from smoke and other environmental stimuli that could trigger later emphysema in children with that genetic makeup. More than half the families suffered severe negative psychological consequences. Consequently, the Swedish government terminated the program two years later.

The identification of a child as having a genetic mutation that predisposes him or her to later illness may also influence how siblings are treated. Dorothy Wertz, Joanna Fanos, and Philip Reilly analyzed how families respond to an ill child, hypothesizing that they might have a similar response to a presymptomatic child. They point out, "Parents are less likely to say, 'When you have children of your own . . . ' to any of their children because they cannot say these words to the ill child." This might impede socialization of the "healthy"/"normal" children.[81]

Existing Recommendations About Testing Children

Because of the potential psychological and financial harm that may be caused by genetic testing of children, a growing number of commentators and advisory bodies are recommending that genetic testing not be undertaken on minor children unless there is an immediate medical benefit to the child. For example, the Institute of Medicine Committee on Assessing Genetic Risks recommended that "in the clinical setting, children generally be tested only for disorders for which a curative or preventive treatment exists and should be instituted at that early stage. Childhood screening is not appropriate for carrier status, untreatable childhood diseases, and late-onset diseases that cannot be prevented or forestalled by early treatment."[82] The International Huntington Association and the World Federation of Neurology recommend that minors not be tested for HD.[83] The National Kidney Foundation recommends that minors not be tested for the gene for adult polycystic kidney disease except in the rare clinical circumstances when measures to prevent stroke are necessary.[84]

In practice, however, these guidelines are not universally followed. Wertz, Fanos, and Reilly note that there is no widely accepted standard among physicians for decisions about genetic testing of children.[85] Despite the fact that the International Huntington Association and the World Federation of Neurology have issued guidelines recommending that minors not be tested for HD, 53 percent of British pediatricians say they would test for the disorder upon parental request.[86] The literature on the subject is also divided. Some commentators suggest that parents should be able to obtain genetic information about children even if there is no potential medical benefit to the child.[87] Others would restrict or prohibit such testing unless there is a medical benefit.[88]

Despite concerns about the genetic testing of children, a survey in the United Kingdom found that many laboratories currently perform genetic testing on children. Some of the laboratories that perform genetic testing on children provide both carrier testing and predictive testing.[89] A British study examining the attitudes of medical professionals toward the genetic testing of children found that 36 percent of participating health care professionals had received requests for predictive testing in children (184 out of 512).[90] Some of the health care professionals indicated that they deferred or refused to perform predictive testing on children for late-onset disorders such as Huntington's disease and polycystic kidney disease. Overall, more pediatricians than geneticists were willing to perform predictive testing on children.[91] While only 4 percent of geneticists would test for Huntington's disease, 38 percent of pediatricians would.

Physicians may test children for various reasons, even when there is no preventive or treatment intervention that can be undertaken, and even when the disease will not manifest until adulthood. Certain physicians have a deep-felt belief in the value of genetic information. Others are swayed by the persuasiveness of marketing.[92] Still others believe (erroneously) that parents have a legal right to such testing.[93]

Rationales for Testing Children

Despite the clear risks to children from genetic testing, parents and physicians give a variety of rationales for such testing. These purported justifications for testing include benefiting a family member, financial planning, recognition of parents' legal rights, and recognition of the state's authority.

In analyzing whether a minor should be tested for the benefit of another family member, Wertz, Fanos, and Reilly use the analogy of organ donation and point out that no one is forced to donate an organ even if it would save the life of a family member. Similarly, children should not be required to participate in family linkage studies or undergo other testing to provide genetic information for other family members.[94]

A joint statement by the American Society of Human Genetics and the American College of Medical Genetics seems to indicate that "social concern" interests may justify testing: "Parents may wish to know about adult onset diseases prior to deciding how much to save for a college education."[95] However, even parents' concern for the financial well-being of a child is not a valid reason to generate predispositional genetic information about the child. In some instances, the child with the mutation will not develop the disease. In others, a treatment for the disease might be discovered before the disease manifests decades later. Even if it were possible to predict that the child will definitely develop a later disease, testing will be unlikely to redound to the child's benefit. There is too much of a chance that such information will cause parents to emotionally or financially neglect a child with a gene for a late-onset disorder.

Parents have great leeway in requesting medical services for their children, but in the past the services have involved the testing and treatment of children who were ill. Questions arise regarding whether parents can also authorize genetic testing for asymptomatic children. It is certainly logical for parents to authorize such testing when there is some preventive or therapeutic intervention that needs to be administered at that age to prevent harm to the child. But should parents also have the right to test children for untreatable disorders—or will that lead to potential stigmatization and discrimination against the child within the family, not to mention the possibility of making the child uninsurable?

Many parents have a strong belief in their right, as parents, to make decisions regarding carrier testing in their children.[96] Forty-eight percent of those parents surveyed who have children associated with Fragile X syndrome felt that they should have the right to decide when their children should be tested and informed of their results. Only 15 percent of parents felt that making the decision to test their child would violate their children's right not to know. Among the reasons given by parents who felt that it was their right were helping their children to adjust to the information, to marry informed, and to have knowledge of their carrier status prior to sexual activity.[97] Ninety-three percent of parents surveyed indicated they were very concerned about their right to make the decision and that they were not concerned that it violated their children's right not to know.[98]

Under common law, parents have a great deal of discretion in making medical decisions on behalf of their minor children. When parents refuse to consent to treatment for their child, pursuant to their constitutional rights of privacy to make child-rearing decisions or to religious freedom, the courts must determine the extent to which state law may legitimately limit parents' constitutional rights. That determination has employed a delicate balance, with a strong presumption in favor of parental rights. The general rule has been to mandate treatment over parental objection only when the child's life is in imminent danger and if the treatment poses little risk of danger in itself.[99] If the minor does not have a life-threatening condition, courts may compel treatment if some permanent affliction will result without it.[100]

Parental rights do not extend to the right to order genetic testing when it could be postponed without adverse health risks until the child is old enough to decide for himself or herself. Hoffman and Wulfsberg note that this is *not* like prenatal testing, where parents have a right to genetic information about their fetus because they could then take a particular action, abortion. "In the context of testing for a genetic predisposition for which the parent can do nothing to alter the likely manifestation of the disease, the physician would not be legally liable. Liability would only attach when a beneficial intervention exists and failure to test or to test in a timely manner would result in harm to the child."[101] Vanderbilt pediatrician and lawyer Ellen Wright Clayton, in an article titled "Removing the Shadow of Law from the Debate About Genetic Testing of Children," additionally points out: "Physicians are required only to provide the same sort of care that would have been provided by other reasonable practitioners . . . and in fact may not perform interventions that may cause their patients more harm than good."[102] She argues that we should debate this issue on the merits—focusing on the effects on children, not some convoluted idea about what the law is.

The fact that some physicians may feel obligated to offer tests means that some parents will feel obligated to authorize testing of their children. They may assume that if a test is offered in a medical setting, it must be beneficial. Such an approach may lead to an increasing number of genetic dossiers being built on children before they reach an age to make medical decisions on their own.

Reconceiving Parenthood

Genetic services have changed the nature of reproduction and parenting. New notions of responsible parenting have surfaced, subjecting decisions made by potential parents to public scrutiny. The experience of pregnancy is dramatically changing. Women are often pressured to use prenatal testing without being informed of the short- and long-term ramifications of such information. In some cases, the only choices available to parents presented with genetic information gleaned from prenatal testing are to carry the fetus to term or to abort. The use of prenatal genetic testing is increasing dramatically, with little attention to the anxiety and emotional turmoil for those who have to make difficult choices during pregnancy, and the impact on the relationship between the family and the fetus or, if the pregnancy goes to term, between the family and the resulting child.

The testing of children for genetic propensities—even in cases where there will be no immediate medical benefit—raises fundamental questions about how far parental rights should extend and what expectations about children are created by genetic information. Policymakers need to explore ways to reduce pressures to use new genetic technologies so that informed decisions about their appropriate uses can be made.

5   The Impact of Genetic Services on Women, People of Color, and Individuals with Disabilities

Genetic services affect people as individuals, but they also affect people as members of groups. Past eugenic practices disproportionally disadvantaged women, people of color, and individuals with disabilities. In trying to create contemporary policies to deal with genetic services, it is useful to understand that history and to analyze the ways in which contemporary health care and genetics practices affect those groups.

Women and Genetic Services

Genetic services do not just affect the individual women who use them. They also shape society's expectations of women in general. Genetic services can exacerbate disparities in health care research and clinical services involving women.

In the past, women have been disadvantaged in research[1] and clinical[2] settings. Their complaints have gone unheard, their diseases have been less likely to be the subject of research than diseases of men, and they have been more likely to be subject to interventions without their consent. In the area of reproduction, women have been subject to risky, surreptitious, and involuntary interventions—not for their own benefit, but for the purported benefit of their offspring.

For countless years, medical research was designed to focus on men's disorders and men's needs. It was not until 1990, once the women in Congress started asking questions, that a Government Accounting Office investigation revealed that women had been systematically excluded from clinical studies,[3] despite a 1986 federal directive to the contrary.[4] It is like the children's story where someone finally points out that the emperor is wearing no clothes. It is now impossible to look at the older studies in the same way again.

It now seems absurd to think that anyone would design a study of the relationship between obesity and breast and uterine cancer, as was done at one institution, that included only men,[5] or that a study of heart disease—the leading cause of death among women—would enroll 22,000 men and no women. It seems equally incomprehensible that a federal study on health and aging would proceed for twenty years with only male subjects.[6] There is even evidence that, in basic research, researchers excluded female rats.[7]

Researchers have also failed to give credence to women's complaints when designing research. For decades, the painful bladder disease of interstitial cystitis was dismissed as a "hysterical female condition."[8] Only through the extensive efforts of a patient advocacy group did the primarily male field of urology begin to pay attention to the disease.[9]

It seems that every relevant entity—the National Institutes of Health, the Food and Drug Administration, institutional review boards, pharmaceutical companies—overlooked what law professor Rebecca Dresser has called this "glaring moral mistake."[10] Even the more recent efforts to include women in research have resulted only in the inclusion of women past reproductive age—ignoring the fact that their hormonal picture is different from that of younger women and their response to medications is different because of the effect of age on the kidneys.[11]

Now the pendulum is swinging the other way. The importance of research on women—including pregnant women—is beginning to be recognized in the policy sphere. Efforts to remedy the situation began in earnest with the National Institutes of Health (NIH) Revitalization Act of 1993, which requires that women be included in intramural and extramural clinical research projects supported by NIH.[12] It also requires that clinical trials be carried out so that valid analyses can be done regarding whether women are affected differently by particular medications or other treatments. Congress specifically intended that these differences be looked at not just statistically but also in terms of their importance and meaning to the women involved.

Various policies are also being considered to prod companies into including informed, consenting women in research, including a potential change in the law to provide extended patent protection for drugs that have been developed through research on pregnant women.[13] Such an approach might create incentives, though, for physicians to pressure or mislead women into enrolling in research studies. A 1995 study found that one in four patients in federally funded research at hospitals did not realize he or she was involved in research.[14]

There are numerous examples of women being "tricked" into certain types of research. In one birth control study, women seeking contraceptives were not told they were part of a research protocol in which some subjects were given a placebo.[15] Ten of the seventy-six women receiving the placebo became pregnant.

In the course of developing reproductive technologies, physicians surreptitiously removed eggs and embryos from women who were undergoing pelvic surgery for other reasons.[16] And apparently Lesley Brown, the first woman to give birth to a child conceived through in vitro fertilization, was not initially told how experimental the procedure actually was.[17] She was led to believe that many women before her had successfully used the technology. Similar deceptions have occurred with each new reproductive technology—such as embryo freezing, egg donation, and egg freezing. Women are led to believe that the technology is well established, even when only a few births based on these technologies have occurred in the world.

Women have also been disadvantaged in the provision of clinical services. Disease itself is defined in male terms. The original definition of AIDS did not recognize the way the disease progresses in women.[18] Consequently, women did not receive adequate treatment—nor were they able to receive Medicaid funding for AIDS care, since such funding was contingent on meeting the male criteria for AIDS.

The health needs of women are often overlooked. Two and a half million women are hospitalized annually for heart disease. Five hundred thousand die annually from heart disease. Despite the fact that heart disease is the leading cause of death in women, women are less likely to be taken seriously when they present with symptoms—and less likely to be referred for an angiogram. While 78 percent of men receive a clot-dissolving medication after a heart attack to prevent future heart attacks, only 55 percent of women do.[19] We have not come that far from the time in 1964 when women who showed up at an American Heart Association Conference on Women and Heart Disease learned to their dismay that it was about how women could care for their husbands' hearts.[20]

Women are much less likely than men to get treatment for many disorders that affect both sexes. Men are 30 percent more likely to get kidney transplants and 65 percent more likely to be referred for cardiac catherization.[21] On the other hand, when a woman gets pregnant, all manner of questionable interventions may be forced upon her, not for her benefit but for the benefit of the fetus.

It is not just in matters of reproduction that medicine has run roughshod over women. Stephen Miles and Allison August found that courts in right-to-die cases have been less willing to honor women's choice to refuse treatment than men's.[22] While the (primarily male) judges respected a man's preference in 75 percent of the cases on withdrawal of treatment, they upheld women's decisions in only 14 percent of cases. Despite the fact that men and women gave similar reasons for refusing treatment, courts characterized men's refusal as "rational" and women's as "emotional" and not sufficiently thought through. In one case, a judge ordered a female Jehovah's Witness to undergo a blood transfusion because she had young children to care for.[23] When a man in that same situation wanted to refuse a transfusion, *his* wish was granted.[24]

Impact of the Earlier Genetics Movement on Women

It is against this backdrop of discriminatory treatment of women in research and clinical settings that the impact of genetic services on women needs to be gauged. The first widespread application of genetics—in the turn-of-the-century eugenics movement—was disproportionately aimed at women. In an extensive analysis of the early eugenics movement and its writings, Nicole Rafter demonstrates how geneticists and policymakers labeled promiscuous women as a social problem and developed institutionalization and sterilization programs to deal with them.[25] An 1879 study asserted, "One of the most important and dangerous causes in the increase of crime, pauperism, and insanity is the unrestrained liberty allowed to vagrant and degraded women."[26] Rafter argues that this approach was actually a mechanism to force women to behave in a socially acceptable way by not allowing them to create children outside of marriage.[27] The institutionalization of the "feebleminded" concentrated mostly on women.[28] Even women who were not feebleminded or promiscuous were targets of legal rules that attempted to assure the health of the next generation. A 1908 U.S.

Supreme Court case, *Muller v. Oregon*, upheld a law limiting the number of hours a woman could work, reasoning that "as healthy mothers are essential to vigorous offspring, the physical well-being of women becomes an object of public interest and care in order to preserve the strength and vigor of the race."[29] Women's challenges to male dominance brought repressive social policies legitimated by theories available in science.

This perspective is typified by the harsh stance taken in the 1927 U.S. Supreme Court case *Buck v. Bell*,[30] a case which has never been explicitly overturned and which was cited with approval in the landmark abortion case *Roe v. Wade*.[31] In *Buck v. Bell*, the Court upheld a statute that allowed the involuntary sterilization of patients of state institutions who suffered from hereditary insanity or mental deficiency. Justice Oliver Wendell Holmes, otherwise a champion of individual rights, wrote the Court's opinion, which authorized the sterilization of Carrie Buck on the grounds that she was feebleminded, stating, "We have seen more than once that the public welfare may call upon the best citizens for their lives. It would be strange if it could not call upon those who already sap the strength of the State for these lesser sacrifices . . . in order to prevent our being swamped with incompetence. It is better for all the world if, instead of waiting to execute degenerate offspring for crime, or to let them starve for their imbecility, society can prevent those who are manifestly unfit from continuing their kind. . . . Three generations of imbeciles is enough."[32] Emphasizing fiscal concerns, Holmes indicated that sterilizing Buck might mean that if she were returned to society she would not be a "menace" but would instead be "self-supporting."[33]

In a stunning research project, historian and lawyer Paul Lombardo showed in 1985 that Carrie Buck, the famous target of Holmes's "three generations of imbeciles" was not an imbecile.[34] She had done well in school, as did her daughter. Rather than being institutionalized because she was feebleminded, she had been institutionalized because she was considered to be "immoral" for having a child out of wedlock.[35] Yet that pregnancy was the result of being raped by the nephew of the foster parents with whom she lived—the very people who committed her to the institution! The doctor who sterilized her was "obsessed with placing checks on sexuality and propagation,"[36] and Buck received appallingly poor legal representation. Her lawyer, who had been on the board of the institution that authorized her sterilization,[37] did not call any witnesses or introduce any facts to challenge the characterization of his client as feebleminded.[38] He obviously was in a position of a conflict of interest and should not have represented her.[39]

Genetic Services Today

Genetic services today also have a disproportionate impact on women.[40] Women are more likely than men to be offered and to undergo genetic testing. Additionally, diagnosis of the fetus more often provides information about the mother than about the father. In the case of a recessive disease, an X-linked disease, and some instances of dominant diseases, the fetus's genetic status will provide information to the mother about *her* genetic status,[41] thus influencing her self-image, her personal relationships, and her relationships with third-party institutions. Women are more likely than men to worry about their results and, if they are carriers of a mutation for a recessive disorder, to think of themselves more negatively.[42]

Compared with men, women are disproportionately targeted for genetic interventions. When the cystic fibrosis gene was identified, some geneticists indicated that it might be useful for couples who were considering having children to be tested, since if both individuals had a mutation of the gene, there was a 25 percent chance that any child they had would receive two copies of the mutated gene and thus would have cystic fibrosis.[43] If one individual was tested, though, and was not found to have an identifiable mutation, then the chance was significantly lower that the couple would have an affected child. The American Society of Human Genetics and a National Institutes of Health Workshop emphasized the need to screen *couples* preconceptually.[44] Either the man or the woman would have the test first and, if the test showed a genetic mutation, his or her partner would be tested as well. However, most clinicians are offering the test not to couples but rather to pregnant women.[45] Since one in 25 whites has a mutation, it means that, when the woman is tested first and only if she has a mutation is her partner tested, then 25 women are tested for every man, even though men and women are equally likely to pass on a mutation to the child. Such an approach disproportionately affects women and may make it seem that genetics is a woman's responsibility rather than a shared one.

Even when a genetic service is properly targeted primarily toward women—such as breast cancer testing—they may not be given adequate information to make a decision about whether to use the service or adequate assurances that the test will be performed correctly.[46] Physicians do not have a particularly good track record for assuring the quality of breast cancer services or for allowing women with breast cancer to make autonomous decisions. The most frequently litigated malpractice negligence claim involves failure to diagnose breast cancer.[47] There has also been litigation by women whose breast cancer biopsies have been misinterpreted, leading to unnecessary surgery and radiation therapy.[48] Moreover, some physicians treating women with breast cancer in the past have not met the proper legal standards of informed consent. In fact, physicians recommending radical mastectomy to women with breast cancer so often failed to meet the legal standard of informed consent that eight states passed laws requiring doctors to inform women of the alternatives to radical mastectomy.[49]

Currently, women may not be given sufficient information about the limitations of testing and preventive strategies. At a meeting I attended, a physician showed a slide saying that if you have a mutation in BRCA1 or BRCA2, you should have yearly mammograms starting at age twenty-five.[50] There was no acknowledgment that such screening may be of limited benefit since the breasts of younger women are more dense, making mammograms less definitive.[51] Nor was there any acknowledgment that the radiation from the mammogram could itself lead to cancer in some women, particularly those with the gene associated with A-T (ataxia telangiectasia).[52]

Some physicians mislead women into genetic testing. For example, a California statute requires physicians to *offer* maternal serum alphafetoprotein screening to all pregnant women. But an observational study of physician-patient interaction by Nancy Press and Carol Browner found that some physicians did not provide adequate information upon which women could make a decision whether or not to undergo the test, and others misled women into thinking that they were required by law to undergo the test.[53]

Other physicians genetically test women without informing them. Some physicians undertake testing on pregnant women without their knowledge or consent, such as testing African American women to see if they are carriers of the sickle cell anemia gene.[54] Physicians do this surreptitious testing on blood collected from the women for other purposes.

Gender-Based Impacts

A person's gender also affects his or her response to genetic testing results. Women perceive themselves to be at higher genetic risk than men. Forty-three percent of women in families with a family history of colorectal cancer feel they have an above-average risk of a gene mutation, in contrast to 30 percent of men.[55] Women are significantly happier than men when they learn they are not carriers of cystic fibrosis and significantly more upset when they learn they are. The Theresa Marteau, Ruth Dundas, and David Axworthy hypothesize that women's greater feeling of responsibility for reproduction leads them to different reactions to threats to reproduction than men experience.[56]

Women experience anxiety when they or their spouses are identified as carriers. The impact of the test results on the spouse is related to the gender of the spouse. Female partners of carriers more often experience heightened anxiety along with their husbands, whereas male partners of carriers less often experience anxiety with their wives.[57] These data indicate that female partners in couples where one or both learn their carrier status are more susceptible to heightened levels of anxiety, regardless of whether they or their partners are found to be carriers.

Women may also be disportionately affected by genetic discrimination. In an analysis of the impact of the Human Genome Project on women, philosopher Mary Mahowald points out that women have long been discriminated against in employment. "The evidence of previous and current employment discrimination based on gender or reproductive potential," she says, "supports the claim that the potential harm of rendering human beings virtually unemployable through genetic prognosis is likely to fall disproportionately on women."[58]

Situations in which genetic testing is undertaken without the individual's knowledge or consent (as sometimes happens to pregnant women) are particularly appalling since women, more than men, feel that doctors should keep out of reproductive decisions. A Swedish study assessing attitudes of women and men toward prenatal diagnosis found that autonomy in the decision-making process was more important to women than to men.[59] In response to the question "Who should decide about prenatal diagnosis, the couple itself or somebody else?" 82 percent of women indicated that the couple should make the decision, but only 20 percent of the male partners gave that response.[60]

It may be more important for women than men to refuse genetic services, since women in general perceive greater risks from technology than men do.[61] Interestingly, nonwhite men are similar to women in that respect. Initially it was suggested that perhaps the greater fear of risks experienced by women and minorities was the result of lesser scientific understanding, but it was found that even well-informed women and minorities shared those perceptions.[62] Fear of technological hazards is the product not just of short-term reactions to the news of some technological disaster[63] but also of systematic differences in the lives of white men versus others. "Perhaps white males see less risk in the world because they create, manage, control, and benefit from so much of it," notes one group of researchers.[64] White women and minority men and women may be concerned about risk because they are more likely to bear those risks than are white men, they have less power and control, and they tend to benefit less from technology.[65]

Genetic Services and the Culture of Motherhood

Today, the extensive use of prenatal testing (with some obstetricians refusing to treat pregnant women unless they agree to undergo such testing) sends the message that it is the duty of women (rather than both parents) to be guarantors of their children's health. When pregnant women do not undergo available prenatal testing, health care professionals blame them for the resulting genetic conditions of their children.[66] Further blame is heaped on women when genetic testing reveals a son to have a condition that was passed on from the woman via the X chromosome—such as Fragile X, Duchenne's muscular dystrophy, or even (as some researchers suggest) homosexuality. Rabbi Elliot N. Dorff has advocated that "women with the defective BRCA1 have a duty to inform their prospective mates of the fact"[67]—apparently so that the men could choose to marry someone else with "better genes." Women may also be criticized for continuing a pregnancy if the fetus is found to have a genetic abnormality.[68]

Such an approach has a negative social impact on women by underscoring the long-standing culture of motherhood that has viewed women as the sole caretakers of their offsprings' well-being. Court cases earlier in this century suggested that women should be forbidden to do certain types of work—including being lawyers—on the grounds that it might make them less fit to reproduce. And when courts upheld sexist employment laws that kept women out of the employment that men were allowed to pursue, they used as a rationale women's childbearing role: "that her physical structure and a proper discharge of her maternal functions—having in view not merely her health, but the well-being of the race—justify legislation to protect her from the greed as well as the passion of man."[69]

The premium put on healthy babies has been seen more recently in cases in which courts have been willing to order cesarean sections for unconsenting women on a doctor's advice that the operation is necessary for the fetus.[70] Psychiatrists have been willing to institutionalize pregnant women who are behaving in a manner considered harmful to the fetus. And legal commentators have proposed statutory systems that would hold a woman guilty of child abuse if she risked harm to the fetus by smoking or drinking during her pregnancy or by refusing to follow doctors' orders.[71]

There appears to be a growing interest in subjecting a woman's pregnancy to public control.[72] Missouri civil rights attorney Arlene Zarembka and University of Arizona law professor Katherine Franke describe this as the "publicization" of pregnancy.[73] Attorney Carol Beth Barnett observes that "once a woman becomes pregnant, her life, her lifestyle and her medical options become subject to public control and scrutiny. . . . From this perspective, a woman's womb is like 'quasi-public territory' and a woman's right to bodily integrity and autonomy receives minimal respect."[74] Such an approach conveys the impression to women and to society that women are mere fetal containers.

Also, as in the earlier eugenics movement, biological explanations are set forth for why women are not suited to become fully functioning members of society.[75] In addition, the concern that poor women contribute to social ills by producing offspring out of wedlock is a common refrain today in the media and in policy discussions.[76] An astonishing 97 percent of obstetricians favor sterilizing unmarried welfare mothers.[77] Several states have proposed legislation that provides incentives for women on welfare not to have additional children. These include offering welfare benefits to women who implant the long-acting contraceptive Norplant (proposed in Connecticut, Florida, Kansas, Louisiana, Mississippi, Ohio, South Carolina, Tennessee, Washington, and West Virginia)[78] or offering women on welfare cash bonuses for undergoing sterilization (proposed in Ohio and Washington).[79] The language used in the current debate on preventing pregnancies in women on welfare sounds like language used in the earlier eugenics movement. West Virginia Supreme Court justice Richard Neely advocates creating incentives for such women to use Norplant: "I am speaking for the Heartland of America, where the underclass is growing by leaps and bounds."[80]

Legal Protections for Women

Only recently have more-enlightened legislatures and judges begun to realize that certain stereotypes about women may cause physicians not to treat women the same way they would men. Consequently, some legislatures have enacted laws to further clarify what "informed consent" means in the context of women's medical decisions, and some courts have upheld the medical decisions of pregnant women even when the fetus might be put at risk.[81]

These policymakers are recognizing that it is improper to claim that women's rights can be overridden in the name of public health in order to get some presumed benefit for the next generation. They recognize, as ethicist Karen Lebacqz notes, "What passes as medical intervention justified on presumed public health grounds cannot be separated from a history of power struggles over who controls women's bodies, women's minds, and women's lives."[82]

A growing number of states have laws that require physicians to disclose information about the nature of, risks of, and specific alternatives to the treatment being proposed, for example, through statutes requiring physicians who treat a woman for breast cancer to describe the alternatives to radical mastectomy. In addition, some statutes require, as part of informed consent, the disclosure of the physician's or clinic's previous success rate with a particular intervention. According to an Office of Technology Assessment study in 1988, one-half of the country's 169 in vitro clinics never had a successful pregnancy.[83] In response, Virginia enacted a statute requiring in vitro fertilization physicians to disclose their success rates to patients.[84] The U.S. Congress similarly mandated that IVF clinics disclose their results to the Centers for Disease Control for public disclosure.[85] Such laws requiring the provision of information have been enacted to curb physicians' tendency to treat women like children and assume that they want the physicians to make decisions for them.

Another legal trend has been to assure that women are not being coerced into accepting a particular medical intervention. Conflicts between physicians and pregnant women have arisen in which physicians have gone to court to force a woman to undergo a blood transfusion, cesarean section, or other intervention that the physician felt would be in the best interest of the fetus. In a 1987 survey of directors of fellowship programs in maternal-fetal medicine, 47 percent were in favor of court-ordered medical interventions during pregnancy.[86] One obstetrician told me he had gone to court fifty times to mandate cesarean sections on unconsenting patients—and judges had ordered forty-nine of the women to submit to the surgeries.[87] But more recent authorities, including medical organization guidelines and court decisions, support the woman's right to make decisions about her pregnancy, labor, and delivery, even in situations in which there might be risk to the fetus.

The turnaround in approach in this area can be traced to a 1990 Washington, D.C., case, *In re A.C.*,[88] in which Angela Carder was admitted to the hospital during her twenty-sixth week of pregnancy with a cancerous lung tumor. Her prognosis was terminal. When she refused a cesarean section, the hospital went to court to get an order requiring the woman to submit. The fetus was surgically delivered and died soon after. Angela died two days later.

The decision and its tragic consequences provoked a storm of controversy and was appealed to a full panel of the appellate court. The D.C. Court of Appeals held that medical treatment decisions for a pregnant patient should be controlled by her wishes, articulated either through her informed consent or, if she is incompetent, through substituted judgment. They should not be made by the doctor. The same result was reached by an Illinois court, which warned against doctors and courts overriding pregnant women's decisions, since it would subject "the woman's every act while pregnant to state scrutiny, thereby intruding upon her rights to privacy and bodily integrity, and her right to control her life."[89]

The American Medical Association, the American College of Obstetricians and Gynecologists, and the American Academy of Pediatrics have all adopted recommendations that recognize a woman's decision-making authority during pregnancy. Also, in recent years, physicians have been found liable for ignoring a woman's refusal of treatment. A Massachusetts woman was awarded $1.53 million in damages for a cesarean performed over her refusal (which resulted in complications and hospitalization for more than a year).[90] Moreover, physicians may be liable for battery if they force medical treatment on a woman, and because battery is viewed legally as an intentional rather than a negligent act, it is unlikely that malpractice insurance will cover it.

Despite the legal precedents protecting women's right to refuse medical interventions, some women are being tested genetically without their consent. At a meeting at an elite medical school, I asked the physicians why they undertook genetic tests on pregnant women's blood without their consent. In that instance, they indicated that they did not ask pregnant African American women for their consent for sickle cell carrier screening using the women's blood because (1) the woman "wouldn't understand," (2) the testing was done for the woman's benefit, and (3) other types of testing are performed without consent during pregnancy. Each of these reasons is open to challenge.

The rationale that the woman would not understand ignores the fact that some people are part of a subpopulation in which genetic disease or genetic testing is common or have relatives with a genetic disorder and therefore do have a high level of understanding about the subject. For example, sickle cell anemia testing has been widely publicized and discussed within the African American community, and it would be unusual for an African American woman not to know someone who had been tested.

On the second point, the fact that the testing is ostensibly done for the woman's benefit does not obviate the need for informed consent. The case law is clear that people have a right to refuse medical interventions, even if those interventions will benefit them.[91] Moreover, it is unclear exactly what the purported "benefit" to the woman is. If the woman is a carrier of sickle cell anemia, she is healthy herself. She has a one-in-four chance of having a child with sickle cell anemia if she reproduces with another carrier, and so it may be a benefit to let her know about the risk in case she would like to have her partner tested or have prenatal testing on the child and abort an affected offspring. But that is far from universally considered a benefit. In some instances, the woman might already know her partner's carrier status. If he is negative for the gene for sickle cell anemia or other hemoglobinopathies, there will be no chance that the fetus will be affected. And even if both partners are carriers and the one-in-four chance materializes and the fetus is affected, the couple most likely will not want to abort. (In a study where women's blood was analyzed for sickle cell carrier status without their consent, *none* of the couples aborted when they later underwent amniocentesis and learned that their fetus was affected.[92]) So it is hard to tell what the "benefit" is.

Related to the argument that this involuntary and unconsented-to testing benefits women is the argument that testing is routinely done on pregnant women without their consent. Putting aside the question of whether *any* intervention should be done in the pregnancy context without the woman's consent, there are reasons why the traditional testing done (such as for gestational diabetes or placenta previa) is distinguishable from genetic testing. Standard, nongenetic tests are often done in order to be able to treat the fetus. Genetic testing often reveals that the fetus is untreatable, so the "benefit" is the possibility of abortion. There is a much wider range of moral and personal opinion about the advisability of abortion than about treatment of fetuses or newborns. Some women understandably may not want information about their genetic status or the fetus's genetic status because they do not intend to abort or because they do not want to risk genetic discrimination against themselves or their future child.[93] Standard tests are also generally for transitional, pregnancy-related conditions, whereas the genetic information revealed about a woman or her fetus is permanent and immutable in character. If a woman learns through unasked-for prenatal testing that she has a genetic defect, from that time on, that information is in her record. Her health insurance rates may go up, or the information may make her uninsurable or unemployable.[94]

Genetic services affect individual women who choose or refuse to undergo testing, but they also have an effect on social expectations for women. They revitalize and reinforce stereotypes of women as guarantors of their offspring's health and change society's expectations about reproduction.

People of Color and Genetic Services

People of color have also been mistreated in the research, clinical, and public health contexts in the past. According to Dr. Jay Katz, author of *Experimentation with Human Beings*, researchers treat members of their own race, class, economic level, and gender much better than they treat people who differ from them with respect to any of those characteristics.[95] This raises concerns for people of color, since most researchers are white males. In 1992, only 3.3 percent of physicians and 2.8 percent of biologists and life scientists were African American (and few of those people were engaged in research).[96] Genetics can exacerbate discrimination. "Racism, prejudice, and genetics have made for a socially combustible and often deadly mix," notes bioethicist Arthur Caplan. "The mixture has proven so toxic that a strong case can be made that applying knowledge from the realm of human genetics to public policy has led to far more misery, confusion, and suffering in the twentieth century than it has to human betterment."[97]

People of color have been subjected to research interventions that were thought to be too dangerous or risky to be used on white people. Slaves were used for extensive, painful medical research. Female slaves were subjected to agonizing, experimental gynecologic surgery without anesthesia, even though anesthesia was available at the time.[98] In an experiment about sunstroke, a male slave was put in an open-pit oven.[99] Thomas Jefferson used an experimental cowpox vaccine on two hundred slaves.[100]

But experimentation on African Americans has continued far longer than slavery. In fact, in the name of medical research, African Americans "have been injected with cancer cells, have had their bodies implanted with plutonium and radium, been severely burned, . . . given experimental vaccines that were unlikely to improve their health, have had parts of their brains cut out, undergone wholesale chromosomal testing for racially related 'criminal predispositions' and have been crudely sterilized with razor blades."[101] African American women have been subjected to experimental abortion techniques that were known to have risks and that had the potential to lead to uncontrollable bleeding, shock, and the need for a hysterectomy.[102]

Public hospitals, which primarily serve people of color, have been the site of research abuses. Between 1986 and 1990, three thousand pregnant women—mostly poor women of color—were given an experimental treatment of steroids by doctors at Tampa General Hospital without their consent. The doctors wanted to see whether the drug would improve lung development of the fetus in pregnancies at a higher risk of delivering prematurely. In 1990, a class action lawsuit was filed on behalf of the pregnant women involved in the study which alleged that the women were not told the drugs used were experimental or that they could refuse to participate. One of the named plaintiffs in the suit underwent amniocentesis eleven times during the last two months of her pregnancy. A decade later, the lawsuit was settled for $3.8 million.[103]

The most well-known—although perhaps not the most infamous—example of research on African Americans is the Tuskegee study. From 1932 to 1972, a United States Public Health Service study of four hundred African American men suffering from syphilis deliberately deprived them of treatment in order to assess the effect of allowing the disease to take its course, even though penicillin had been found to be an effective treatment.[104] In May 1997 President Clinton offered a formal apology for this gross mistreatment, saying, "To our African-American citizens, I am sorry your federal government orchestrated a study so clearly racist."[105]

In addition to serving as guinea pigs for research on medical therapies that could later be used beneficially on whites, people of color have been subjected to research specifically designed to demonstrate their inferiority. Studies have been undertaken measuring African Americans' and Native Americans' skulls and brains to "demonstrate" their lesser cognitive ability.[106]

More interventionist "treatments" have been proposed for primarily African American groups as well. In the 1970s, U.S. psychosurgeons proposed to treat inner-city violence by lobotomizing ghetto militants.[107]

Even today, in the clinical setting, studies show that African Americans are less likely to get necessary surgical procedures than whites; the results are the same even after controlling for socioeconomic and clinical factors.[108] Even when their medical conditions are similar to those of whites, African Americans receive fewer appendectomies, cardiac valve replacements, lumbar disk procedures, and tonsillectomies.[109] The only category in which they received more procedures than whites was operations that relate to their reproductive organs (hysterectomies and prostatectomies), which further underscores stereotypes about race, sexuality, and reproduction.

African American women are 2.2 times as likely to die from breast cancer than white women, yet only 42 percent are diagnosed at the early stage, compared to more than 50 percent of white women.[110] Even contemporary researchers who are trying to understand the differences may not be sufficiently racially sensitive. The NAACP has denounced a Memorial Sloan-Kettering Cancer Center survey that asks African American women if they eat chitlins, believe in voodoo, or seek care from root doctors.[111] "What do chitlins have to do with breast cancer? It is insulting, it is insensitive, it is provocative, and it is stereotyping," says Marshall England, the chairman of the Harlem Hospital community advisory board.[112]

Impact of Earlier Genetics Laws on People of Color

People of color have been subjected to eugenics laws in the name of public health. Starting with Connecticut in 1895,[113] thirty-four states enacted antimiscegenation laws, prohibiting interracial marriages (which were thought to lead to defective offspring).[114] Even programs designed to benefit African Americans have worked to their disadvantage. In the early 1970s states adopted laws mandating sickle cell carrier status screening of African Americans. The programs, however, lacked a counseling component and provided inadequate protection for confidentiality.[115] Since the purpose of the laws was to help African Americans change their reproductive behavior to avoid having children with sickle cell anemia, they were criticized by some African Americans as being genocidal.[116]

Genetic information has been used as an excuse for discrimination against African Americans. At the time of the mandatory sickle cell screening laws, University of Chicago pathologist James Bowman uncovered various instances of unfounded discrimination. African American airline employees who were sickle cell carriers were grounded, and insurers charged higher rates to sickle cell carriers even though there was no evidence that such people (who had one gene with the mutation for the disorder, rather than the two genes necessary for sickle cell anemia), were at a higher risk of illness or death.[117] Only after extensive lobbying were the mandatory sickle cell screening programs repealed or replaced by voluntary screening programs.

Herbert Nickens notes that the sickle cell experience shows how "genetic disease can serve both as a rallying point for drawing attention to the needs of a group, and at the same time a source of increased stigmatization with imputation of innate defectiveness."[118] Since minority status itself carries a certain stigma,[119] adding the mark of a genetic mutation can create a double stigma. "Because of the stigma associated with minority status, any disvalued trait associated with minorities has that negative valence amplified," writes Nickens.[120]

Impact of Genetics Research on Minority Individuals

Minority individuals have not fared well in the research, clinical, and public health application of genetics. Much of the field of molecular biology has used cells from the HeLa cell line for research. Those cells were cultured from a thirty-one-year-old African American woman, Henrietta Lacks, who died of cervical cancer in 1951. Even today, researchers can purchase cells from her cell line. Virtually every biologist has worked with those cells. But Henrietta Lacks's cells were used without her permission or that of her family. Her husband considers this to be "exploitation."[121]

Today, in the research context, money is disproportionately spent on studies of genetic diseases affecting whites. In 1992, when the NIH cystic fibrosis budget was $46 million (with an additional $18 million in support provided by the Cystic Fibrosis Foundation), the NIH allocation for sickle cell anemia research was $18 million, even though there are more individuals in the United States with sickle cell anemia than cystic fibrosis,[122] and even though both diseases affect people in childhood and kill many people before middle age.

In the Human Genome Diversity Project, researchers are scouring the globe for indigenous groups who may have genes that are useful to whites in developed nations. In March 1995 researchers from the National Institutes of Health (NIH) obtained a patent on the DNA of a man from New Guinea whose genes protect him from leukemia.[123] Similarly, the Wellcome Foundation has used the Nawalwa cell line, taken from an African child with lymph cell cancer, to produce a valuable pharmaceutical, interferon. Some critics of the Human Genome Diversity Project describe such practices as "biopiracy" and "biocolonialism." They note that it is unlikely that the Third World peoples whose bodies are being used as a pharmaceutical resource will be able to afford the resulting products. In 1993 the World Council of Indigenous Peoples unanimously voted to "categorically reject and condemn the Human Genome Diversity Project as it applies to our rights, lives, and dignity."[124] They point out that they live in despicable conditions, deprived of traditional rights and land, and are housed together on reservations. While they are living in poverty, their cells are creating profits for someone else. As one representative of an indigenous group opined, "You've taken our land, our language, our culture, and even our children. Are you now saying you want to take part of our bodies as well?"

And, as has happened historically, genetic research is being used to "demonstrate" the genetic inferiority of people of color. *The Bell Curve* asserts that African American individuals, as a group, have lesser mental capacities than white individuals.[125] This assertion is used as a rationale to deny funding to programs such as Head Start or college affirmative action initiatives, on the grounds that African Americans' potential achievements are biologically limited. In another sphere, inner-city youths are alleged to have genetic differences that cause them to be violent, and pharmacological intervention or gene therapy is proposed to change their behavior.[126] This kind of reasoning is reminiscent of a 1970 study funded by the National Institutes of Health in which blood samples were taken from more than 7,000 boys, more than 95 percent of whom were from underprivileged African American families, so that their blood could be tested to see if they had an extra Y chromosome. There was a theory—later disproven—that XYY males were more likely to become violent criminals than males without the extra Y chromosome. Another 6,000 young men (85 percent of whom were African Americans) in state institutions for abandoned or delinquent children were similarly tested, and the genetic test results were routinely passed to courts to use however they chose.[127]

There is reason to be skeptical of assertions that people of color are genetically inferior, which, as a century ago, fits so comfortably with current social ideologies. Sociologist Dorothy Nelkin notes, "Behavioral genetics is in vogue these days—just as eugenics was in the 1920's—in part because it suits the political context, providing justification for social policies and legitimation for political goals."[128] Genetics researcher Alan Tobin adds, "Predestination (of whatever sort) means that no one is responsible for inequalities in the society, for the success of some and the failures of others."[129]

Genetics and Race in the Clinical Setting

In the clinical setting, women of higher socioeconomic groups are able to make greater use of prenatal screening; this may lead to a higher proportion of children with disabilities being born to minority women.[130] Even when they gain access to testing, minority women are often treated differently than white women. While genetic testing of pregnant white women for the cystic fibrosis mutation is undertaken with elaborate consent procedures,[131] pregnant African American women are tested for the sickle cell mutation without their advance knowledge or consent.[132] Yet the power to refuse interventions may be particularly important to African Americans, given their past history of powerlessness, and they may believe that their own individual and cultural values are furthered by such refusals. Some African American women refuse amniocentesis, for example, because of concerns about what other uses might be made of the tissue.[133] The National Marrow Program has difficulty recruiting black donors because of concerns that their marrow will go to white patients or that the genetic information collected in the course of donation will be misused.[134] Furthermore, African American women's distrust in the medical profession makes them less interested than other women in undergoing breast cancer genetic testing.[135]

Stigmatization of People of Color Through Testing

Differentially subjecting people of color to testing stigmatizes them as inferior. Along those lines, the Lawrence Berkeley Laboratories ran a program in which they tested African American employees for sickle cell carrier status, and the test results were put in the employment files but were not disclosed to the employees. Whites were not tested for any genetic disorders. Both African American and white employees were tested surreptitiously for syphilis, but African Americans were tested more frequently. Such a testing program reinforces inappropriate stereotypes about African Americans' being genetically inferior and having unchecked sexuality. Thomas Budinger, the former medical director of the lab, underscored the stereotype. When asked why minorities were singled out for repeated syphilis testing, he said, "Because that's where the prevalence of the disease is."[136]

A forty-six-year-old administrative assistant at the lab, Vertis Ellis, learned that at each of her six company physicals over her twenty-nine years of employment, she had been tested without her knowledge or consent for sickle cell anemia, syphilis, and pregnancy. Ellis says, "I felt so violated. I thought 'Oh my God. Do they think all black women are nasty and sleep around?'"[137]

Mark Covington, another lab administrative assistant, tested positive for sickle cell carrier status, but was not told of the results. "It's disgusting," he says. "It goes back to a time when blacks were treated like filthy animals, and they wanted to make sure we weren't contaminating the environment."[138] Despite these concerns, a federal trial court held that the program (which used blood samples provided for other purposes during employee physicals) did not invade the employees' privacy. A subsequent appellate decision, however, sided with the workers.[139] The court held that differentially testing African Americans not only invades their privacy but also is a form of race discrimination.

Genetics and Minority Group Membership

Genetics will affect people of color in even more ways in the future, as geneticists make claims that they can determine race and ethnic status based on genetics. Researchers now assert that they can distinguish between blacks and whites based on differences in just three of a person's 100,000 genes.[140] Native Americans are also thought to be identifiable based on a small number of genetic markers.[141]

In the past, minority groups were penalized because they were thought to have inferior genes. Various social programs, such as affirmative action, have developed as a means of compensating for those past abuses. Ironically, however, people who have suffered discrimination based on their purported inferiority may be excluded from recompense on the grounds that their genes indicated that the discrimination they suffered was erroneous. A woman who looks like a Native American, who has been raised on a reservation as part of a tribe, might in the future be denied a scholarship because her genetic profile does not match the one that researchers claim identifies Native Americans.

Genetic data might be used to take land back from Native American groups on the grounds that they are not Native American enough.[142]

In his essay "Handle with Care: Race, Class, and Genetics," bioethicist Arthur Caplan asks, "Will the information generated by the genome project be used to draw new, more 'precise' boundaries concerning membership in existing groups? Will individuals who have tried to break their ties with ethnic or racial groups be forced to confront their biological ancestry and lineage in ways that clash with their own self-perception and the lives they have built with others?"[143]

As genetic determinism filters through society, its chance to damage people of color grows. Genetic research may be used to "justify" existing social inequities. It can be used to assign people to places of genetic inferiority. And, in a particularly vexing paradox, genetics can be used to deny people minority-group membership in ways that cause a loss of social identity and social benefits.

Individuals with Disabilities and Genetic Services

As the experience of women and minorities demonstrates, medical research and clinical services have been applied in disturbing ways to marginalized individuals. People with disabilities, in particular, have been targeted. In large measure, the history of eugenics is a history of brutality against the disabled. People who were mentally disabled were involuntarily sterilized in the United States—by the thousands—at the turn of the century. In Nazi Germany, people with disabilities were systematically exterminated. Even today, much of the writing about genetic discoveries includes economic analyses about the cost of care for people with a particular genetic mutation, implying that society would be better off had they not been born.[144]

This is a personal matter for Marsha Saxton, a disability rights activist and lecturer at the University of California, Berkeley, who was born with spina bifida. "When I sit in meetings with state public health officials who are showing slides indicating how much the state could save if pregnant women aborted fetuses with my condition," she says, "I have to keep saying to myself, 'They don't want to kill *me*.'"

Genetic technologies are having a major impact on people with disabilities. First, genetic tests create new categories of the disabled by identifying asymptomatic people as being likely to suffer from later diseases. Second, the availability of such testing is creating pressure on parents to avoid having children with disabilities; it allows parents to identify an increasing number of less serious disorders before birth and to terminate affected embryos or fetuses. Third, genetic tests stigmatize existing people with disabilities as having slipped through the net of prenatal screening. The mere existence— and marketing—of genetic testing may seem like an affront to many people with disabilities, given the numerous oppressive ways in which "genetic" information has been used in the past, as with the mandatory sterilization laws at the turn of the century in this country and the ways in which disabled individuals continue to be discriminated against today.[145] The fact that the birth of children with certain disabilities can be "prevented" can aggravate the stigmatization of such children if they are born.

"White society may discriminate against black adults or children, but, historically, black women and children could count on care, love, support, and a modicum of respect within their own families and communities," notes Adrienne Asch. "Not so for the woman with a disability or for most children with impairments. The lack of a natural communal or familial structure can be psychologically and socially devastating."[146] Some family members may treat people with a disability as harshly as outsiders do.

Geneticist Angus Clarke points out that few of the health care professionals who offer prenatal diagnosis and abortion for particular conditions actually treat individuals with those conditions. "Whatever our personal feelings, our lack of involvement in such work must convey the impression that we think it less important than prenatal testing and secondary prevention," he says.[147]

Changing Concepts of Normality and Disability

Genetic testing changes the very categories of "disabled." As bioethicist Paul Ramsey pointed out when amniocentesis was first introduced, "the concept of 'normality' sufficient to make life worth living is bound to be 'upgraded' "[148] as testing is increasingly offered for less and less serious disorders. Currently, some parents choose to abort for reasons that seem trivial or inappropriate to other people. Some parents abort fetuses with an XYY chromosomal complement, for example, even though research has disproved the hypothesis that this is a "criminal" genetic profile. People with dwarfism "are incensed by the idea that a woman or couple would choose to abort simply because the fetus would become a dwarf."[149] As time passes, an increasing number of women abort based on particular types of genetic information, which also changes the boundaries of normality. The percentage of women in Scotland who aborted fetuses with spina bifida rose from 21 percent in 1976 to 74 percent in 1985.[150] A majority of genetic counselors (63 percent) say they themselves would abort a fetus with Down syndrome.[151]

Newer genetic technologies provide the opportunity for users to choose to avoid very minor deviations from some perceived normality. Consider the use of pre-implantation screening, which Marsha Saxton refers to as "admission standards" for fetuses.[152] A couple undergoes in vitro fertilization, produces multiple embryos, and each one is genetically tested. If a couple has ten embryos in petri dishes, they might use different criteria to determine the "genetic worth" of the embryos to be implanted than they would in determining whether a single in utero fetus at five months of development should be aborted. While a couple may not be likely to abort a fetus based on its sex or based on its being an unaffected carrier of a recessive disorder, they may, when faced with a high volume of embryos, only a few of which can be safely implanted in the woman, decide to implant only noncarrier embryos or embryos of a particular sex. With ten embryos, the couple must refuse implantation of some of the embryos—for the safety of both the woman and any resulting fetuses. The choice to move from randomly selecting the embryos to be implanted to selecting them on genetic grounds may seem less morally problematic to the couple than aborting a particular fetus. The decision may be viewed differently for several reasons. The woman is not physically pregnant, so there has not been attachment to a particular fetus. She has not felt the fetus move or begun to bond with it. In addition, the process may be viewed less as choosing *against* a particular individual (the developing fetus) than as choosing *for* a set of individuals (the embryos to be implanted).

Physicians may also try to have greater influence on a woman's decision when the embryo is in *their* laboratory rather than *her* womb. If, for example, the couple have two embryos, both affected with Fragile X, and ask the doctor to implant them, he may urge that they try again or use a donor gamete.[153] Bioethicists Heather Draper and Ruth Chadwick observe that when women's embryos are tested, "it is possible for them to lose control over what happens next." The woman "cannot compel him [the doctor] to implant embryos against his wishes."[154]

Some genetic testing programs go even further than reporting known conditions; they also report any unusual genetic pattern, even where there is no clear indication about whether the genetic deviation has *any* health implications whatsoever. Thus couples may choose not to implant embryos and not to carry to term fetuses that do not meet some gold standard of normality, even in instances where the fetus would suffer absolutely no health risk.

In addition to possibly making even less serious departures from the model genome seem like a disability, genetic predispositional testing is creating a new form of disability in which currently "able" individuals are treated by insurers, employers, and others as "disabled" because of a potentially increased future propensity to illness.

Women who undergo genetic testing on themselves or their fetuses and people with disabilities have much in common. Both groups may suffer from medical manipulation. A woman whose genetic test results indicate that her fetus is affected with a genetic disorder—or that she herself is likely to develop a genetic disorder later in life—may be discriminated against by various social institutions such as insurers or employers. Yet, despite certain commonalities, these two groups may appear to be at odds with each other, for some women's use of prenatal diagnosis to identify fetuses with particular disorders and abort them may be viewed as devaluing existing people who have those same disorders. In some—perhaps many—instances, the decision may be based on ignorance or misinformation about what life with that disability is like, or even whether people with the particular condition consider it a disability at all.

Pressure to Use Genetic Services to Eliminate Disability

"Prenatal screening seems to give women more power," says disability rights activist Laura Hershey, "but is it actually asking women to ratify social prejudice through their reproductive 'choice'?"[155]

Genetic testing creates an environment in which people believe that if a test is offered they should take it and act upon the information. "Women are increasingly pressured to use prenatal testing by claims that undergoing these tests is the 'responsible thing to do.' Strangers in the supermarket, even characters in TV sit-coms, readily ask a woman with a pregnant belly, 'Did you get your amnio?'" notes Martha Saxton.[156] A government agency, the Office of Technology Assessment of the U.S. Congress, exemplified this approach. After describing new genetic tests, an Office of Technology Assessment report stated, "Individuals have a paramount right to be born with a normal, adequate hereditary endowment."[157] Similarly, the report of an NIH task force on prenatal diagnosis states: "There is something profoundly troubling about allowing the birth of an infant who is known in advance to suffer from some serious disease or defect."[158]

The way in which physicians describe a genetic condition may make it seem much more grim than it seems to a person with that condition. "Medical descriptions of Down syndrome—rather than revealing the variability of the condition—selectively represent the condition in uniform, distancing, negative, ungendered, and static terms," notes Diane Beeson.[159] In the prenatal setting, as an increasing number of tests are being developed for less serious disorders, parents may be pressured into feeling that something is "wrong" or "unfit" with their baby if there is the slightest departure from a socially conditioned perfection. Prenatal diagnosis, asserts McGill University epidemiology professor Abby Lippman, is "an assembly line approach to the products of conception, separating out those we wish to develop from those we wish to discontinue."[160]

Genetic testing may also "privatize" disability. Once prenatal diagnosis and testing are made available for a particular disorder, there may be a tendency to discontinue funding for research to help combat the medical problems for existing people with that disorder and to discontinue social services for such individuals. Once genetic disease is no longer seen as a random characteristic, our society may reduce its communal commitment to people with genetic disabilities.[161] This may be especially true in the coming years as wealthier women get access to prenatal diagnosis and abortion and poorer women do not.[162] If a disproportionate number of disadvantaged women have children with disabilities, their lack of clout in state legislatures may make it less likely that legal protections such as educational opportunities for people with disabilities will be continued.

Currently, the law creates a duty on the part of physicians to offer genetic services to high-risk patients. Since women have a constitutional right to abort, they have a right to know about their increased genetic risks and the availability of genetic testing so they can either avoid conceiving a child with a particular genetic disorder or terminate a pregnancy with an affected fetus. Most states recognize wrongful-birth cases, allowing couples to sue if their physician failed to inform them about genetic testing or failed to perform the tests correctly and they give birth to an affected child. The courts allow such suits only if the disorder is "serious." One court indicated, for example, that deafness was not sufficiently serious. However, parents have won cases in which the resulting child had Down syndrome, cystic fibrosis, and polycystic kidney disease.

Courts have been much less willing to recognize wrongful-life suits on behalf of the child, rather than on behalf of the parents. The claim of the child in a wrongful-life case is that he or she would rather not have been born than be born with a particular disorder—and most courts have believed that to be a matter for philosophers, rather than judges, to decide.[163]

The courts' willingness to recognize wrongful-birth suits, though, contrasts greatly with decisions in the 1960s, when courts did not allow such suits, in part because they would devalue people with disabilities. In *Gleitman v. Cosgrove*,[164] a 1967 case, for example, the court stated:

> The right to life is inalienable in our society. A court cannot say what defects should prevent an embryo from being allowed life such that denial of the opportunity to terminate the existence of a defective child in embryo can support a cause for action. Examples of famous persons who have had great achievement despite physical defects come readily to mind, and many of us can think of examples close to home. A child need not be perfect to have a worthwhile life.
>
> We are not faced here with the necessity of balancing the mother's life against that of her child. The sanctity of the single human life is the decisive factor in this suit in tort. Eugenic considerations are not controlling. We are not talking here about the breeding of prize cattle. It may have been easier for the mother and less expensive for the father to have terminated the life of the child while he was an embryo, but these alleged detriments cannot stand against the preciousness of the single human life to support a remedy in tort.

The court even went so far as to cite Jonathan Swift's "A Modest Proposal"[165] to support the last point.

"There is reason for us to fear wrongful birth suits and oppose suits for wrongful life," says Adrienne Asch, a blind disability rights activist and professor at Wellesley College. "It is the message they send to the children themselves, disabled people, and society about the worth of life with impairments."[166]

Preventing the birth of an individual with a disease is morally different than preventing a disease. "It suggests that the lives of some persons with a disability or illness are not worth living, that such persons are to be understood only as social or economic drains and never as sources of either independent value or enrichment for the lives of others," says Ruth Faden, director of the Program in Law, Ethics, and Health at Johns Hopkins University.[167]

Couples may be made to feel guilty by relatives, health care providers, insurers, and other social institutions if they give birth to a child whose departure from "normality" could have been determined prenatally. This is particularly true in the case of mothers who already have children with a genetic disorder and who give birth to additional children with the same disorder.[168] Couples may be the target of hostility for carrying a child with even a slight disability to term, as occurred with Bree Walker-Lampley's decision to give birth to a child with ectrodactyly, a mild genetic condition that fuses the bones in the hands.[169]

Ruth Faden argues against this new requirement of responsible mothering. "Maternal commitments to care for one's children, to seek their interests, and to spare them disease and disability . . . do not extend to or encompass preventing the birth of such children (whether by selective abortion, selective conception, or remaining childless)."[170]

Protecting the Right to Refuse Testing

It used to be that most people who were offered genetic testing had some family history of the disease. They had a sibling, or cousin, or other family members with the disorder. Or, in the case of prenatal testing offered because of advanced maternal age, they knew that the purpose was to identify fetuses with Down syndrome, and they probably had at least met someone over the course of their lives with a child who had that condition.

Testing now covers such a wide range of conditions and disorders that the women being offered prenatal testing—and even some of the obstetricians ordering the tests—have no idea what the life of a child with such a genetic makeup would be like. They must increasingly depend on the people marketing the tests, such as biotechnology companies, for information about how serious the disorder being tested for might be. In the context of breast cancer genetic testing, a biotechnology company exaggerated the risk of cancer that women with the genetic mutation faced. Such tactics can push people into undergoing testing who otherwise would not have. Moreover, decisions may be made about terminating fetuses based on certain stereotypes about disability. Asch points to research studies that show that "whites and middle-class people in general showed more discomfort with Down syndrome and retardation, whereas people of color and those of lower socioeconomic status expressed more fear of physical vulnerability."[171]

Relying on physicians' assessments of disability can be problematic. Even those physicians who treat people with disabilities may have inaccurate impressions of the lives of such people if the physicians interact with them only in a medical setting.

Physicians have a much different view of particular disabilities than do people with those disabilities. When asked to evaluate the quality of their lives, 80 percent of doctors said pretty good, but 82 percent indicated that their quality of life would be pretty low if they had quadriplegia. In contrast, 80 percent of people with quadriplegia rated the quality of their lives as pretty good.[172] Consequently, as Adrienne Asch indicates, "those wishing to use the [genetic] technology should receive substantially more information about life with disability than they now do."

Laura Hershey is a poet and newspaper columnist. "I have a rare neuromuscular condition," she says. "I rely on a motorized wheelchair for mobility, a voice-activated computer for my writing, and the assistance of Medicaid-funded attendants for daily needs—dressing, bathing, eating, going to the bathroom."[173]

"My life of disability has not been easy or carefree," she continues. "But in measuring the quality of my life, other factors—education, friends, and meaningful work, for example—have been decisive. If I were asked for an opinion on whether to bring a child into the world, knowing she would have the same limitations and opportunities I have had, I would not hesitate to say, 'Yes.'"[174]

A 1999 British study by RADAR (the Royal Association for Disability and Rehabilitation) surveyed people with disabilities and predispositions to disabilities. Many disapproved of the current approach to prenatal diagnosis. Thirty-seven percent would ban abortion for Down syndrome; 54 percent for a painful condition that didn't manifest until age forty; 56 percent for deafness; and 72 percent for a correctable genetic condition such as cleft palate.[175]

Other people note the strange contradiction that just at the political moment when laws such as the Americans with Disabilities Act are enacted to protect people with disabilities, genetic technologies are aimed at preventing their birth. "It is ironic," says Marsha Saxton, "that just when disabled citizens have achieved so much, the new reproductive and genetic technologies are promising to eliminate their kind—people with Down Syndrome, spina bifida, muscular dystrophy, sickle cell anemia and hundreds of other conditions."[176]

Disability rights activist Lisa Blumberg asserts, "Too often counselors do little more than provide future parents with a dreary laundry list of problems their child could have and express sympathy."[177] Before testing, an individual should be told about the genetic conditions for which he or she or the fetus is being tested, whether they are treatable, and whether, in the prenatal context, the parents will be faced with a decision about aborting the fetus. He or she should have the opportunity to meet people with the disability for which testing is being done. Much thought should be given to the type of information that is presented; in this area, disability activists can provide considerable aid to physicians and counselors.

Some people with disabilities do support prenatal diagnosis and pregnancy termination. Geneticist Angus Clarke argues that they may do so at least partly because their own lives have been blighted by social, as opposed to medical, factors. He makes the analogy to elderly people who might be induced to undergo assisted suicide, though they would prefer to live, because they want to cease being a "burden."[178] Adrienne Asch sees the issue differently. She views the disability rights movement as pro-choice and thus as having a dilemma with advocating against a woman's right to abort in any circumstance.[179] Providing more accurate information about an individual's life with disabilities and assuring that people in general have more contact with such individuals in the course of their daily lives[180] may help prevent some couples from being coerced into testing and abortion. Personal knowledge of a particular disorder can diminish the tendency of people to unthinkingly seek a "genetic fix." Some genetic counselors, for example, find that after working with patients with cystic fibrosis, they are no longer willing to participate in counseling for prenatal tests (since they personally do not feel that the birth of such a child should be prevented). In a study of parents of children with cystic fibrosis, only 20 percent said they would abort for that disorder.[181]

Who should make the determination about which conditions are the proper subject of genetic testing? Some commentators advocate banning all (or certain) types of genetic testing as being antithetical to a respect for people with disabilities. Others discuss how, if genetic testing is undertaken, to do it with sufficient respect for its effects on women, people of color, and people with disabilities.

The push to test—even for trivial or treatable conditions—is strong, in part because there is a financial benefit to the companies, patent holders, and physicians who offer testing for less serious disorders. Yet perhaps certain tests should not be offered—because the condition is so trivial, is treatable after birth, or the existence of the test seems to demean already-existing individuals with that condition. We discourage prenatal sex selection, because it is assumed that fetuses should not be aborted just because the individual will be discriminated against later in life. As we are flooded with new genetic tests, we need a society-wide assessment about where to draw the line on the use of genetic services.

Protecting the Disadvantaged

Modern genetic techologies offer an opportunity to further extend the elaborate discriminatory history in all areas of life, including education, employment, social status, and medical treatment. Genetic arguments were part of the social Darwinism earlier this century that insisted that those on the bottom of the social ladder (women, minorities, and those with disabilities) belonged there—indeed, should affirmatively be kept there if they attempted to claim a higher place. Today, biological explanations are again given for the inferiority of the least powerful. And, again, women, minority groups, and people with disabilities are targets of the concern raised about the demise of the gene pool. They are also the groups most likely to have their individual decisions overridden, sometimes on the patronizing grounds that it is for their own good, other times for the supposed good of society. Allowing individuals in disadvantaged groups to make informed, more autonomous choices could reduce the resulting stigmatization and inequalities. Because genetic services can lead to further labeling and to the sorts of social expectations that have stereotyped and limited certain groups in the past, there is a special need to examine the differential impact of these services on the disadvantaged.

6 Problems in the Delivery of Genetic Services

Genetic tests and treatments are so rapidly entering the clinical arena that physicians may not have sufficient expertise to assess their worth. In addition, certain unique features of genetic testing make it more difficult to assure the quality of genetic services, yet patients and policymakers may not be aware of that fact. And the close ties between biotechnology companies and university and government genetics researchers mean that there are few scientists in this field who do not have a financial conflict of interest when it comes to assessing the validity of and need for genetic services.

What are the appropriate uses of genetic services? For most medical services, individuals rely on their physicians for advice, and society relies on the scientific community for assessments of the risks and benefits of those services. This paradigm does not necessarily work well for genetic services, however. The commercial push to introduce tests has led to testing being used prematurely. In addition, since genetic services in the reproductive context may involve termination of affected fetuses, physicians' personal moral views may color their supposedly neutral medical advice. Physicians are accustomed to being directive about use of their services, which may not be appropriate with presymptomatic testing, where the psychological and social risks may outweigh the benefits. Additionally, in the prenatal setting, physicians and patients may not necessarily share the same views about the appropriate circumstances, if any, for pregnancy termination.

Physician Knowledge About Genetics

There are problems with relying on physicians to disseminate new genetic information. Not all medical schools offer courses in genetics. A 1995 survey found that only 68 percent of 125 American medical schools required students to take a genetics course.[1] In some of those schools, the "course" was only four hours long. According to an interdisciplinary federal advisory task force, "genetics is not being taught adequately to all medical students"; the group recommended that the schools' "clinical departments pay greater attention to genetic issues."[2]

Family practitioners may be the primary source of genetic information for families, yet there are gaps in their knowledge. Family practice residency programs generally do not include genetics; a poll of seventy-four such programs found that none included clinical genetics components. On tests of genetics knowledge, the average score was 53 percent for faculty and 42 percent for residents in a family practice residency program. The researchers then reviewed charts of patients with genetic disorders and found that an appropriate clinical response was indicated in only one-third of the cases.[3]

Because of the explosion of genetic tests, physicians may not be adequately trained or motivated to understand the nature of certain genetic diseases or the medical and psychological implications of testing. In a study of obstetricians and family practitioners with delivery privileges in Rochester, New York, 35 percent incorrectly believed one needed to have an affected relative in order to have a child with cystic fibrosis, and 43 percent believed that cystic fibrosis affected children of only one sex.[4] In a national survey conducted by researchers at Johns Hopkins, 99 percent of geneticists knew that if only one parent was a carrier, there was almost no chance of having a child with cystic fibrosis, but only 55.7 percent of obstetricians/gynecologists knew that.[5] In one study, 20 percent of people who were carriers of sickle cell anemia erroneously thought they themselves would have health problems.[6] They received that misinformation from doctors!

Many new genes are discovered each month and move rapidly into use in clinical diagnosis without physicians being adequately prepared to advise patients about the appropriate circumstances for their use and the interpretation of results. For example, in 1996 more than 20 percent of 124 primary-care physicians responding to a survey on testing for genetic susceptibility to cancer were unaware of any test for a genetic predisposition to breast cancer.[7] In an assessment of the genetics knowledge of more than 1,000 primary-care physicians, the average number of correct responses was only 74 percent.[8] In another study, residents in obstetrics and gynecology scored between 54 percent and 65 percent correct on questions about clinical genetics and genetic testing.[9]

A study showing that one-third of physicians erroneously interpreted the results of genetic testing for colorectal cancer points to quality assurance issues in another aspect of genetic testing, the interpretation of results.[10] The probabilistic information that is generated by genetic testing—indicating that the patient has an increased percentage chance of developing a disease—is particularly difficult for physicians to deal with. They are accustomed to dealing with patients who are ill, so they tend to overestimate the chances that people with a genetic mutation will actually later manifest the disease. Thus, sometimes their patients get—and possibly act on—genetic information that may be meaningless. For example, with breast cancer genetic testing, Myriad Genetics reports mutations to patients even if there is no indication that the particular mutation is correlated with a higher incidence of cancer.[11] Sometimes "mutations" are actually benign variants of the normal gene; without epidemiological studies to determine correlation, the information about some mutations is virtually meaningless. Even in instances in which those correlations have been made, not everyone with the mutation will necessarily get the disease. Half of the women with the 185delAG breast cancer gene mutation will not get breast cancer. Yet physicians often treat all women with the mutation as if they will get the disease, subjecting many women to intrusive and unnecessary procedures.

The fact that many genetic diseases do not have a high incidence also creates problems with respect to physician advice. If a disease has a one in 15,000 incidence and the test for it is 99.9 percent accurate, with a 0.1 percent false-positive rate, that means there will be fifteen false positives for each true positive. Those fifteen people may receive unnecessary and possibly dangerous treatment. They may also suffer the emotional stress of thinking they have a genetic disorder when they actually do not.

Studies by Johns Hopkins School of Medicine professor Neil Holtzman show how poorly physicians understand the sort of probabilistic data that the new genetics raises. He found that genetic reasoning deteriorates during medical school. He asked, "If a test to detect a disease that occurs in one of 1,000 people is falsely positive in 5 percent of unaffected people, what is the chance that a person found to have a positive result actually had the disease?"

Eighty-four percent of first-year students, 73 percent of third-year students, and 49 percent of interns and residents selected the correct answer, "2 percent." The physicians who gave erroneous answers overestimated the predictive value of the test—they were ready to believe a test result and begin treatment on the basis of it, even when the patient was fifty times more likely to be healthy than diseased. This result was disheartening, given that these were the actual odds of some widely used genetic tests, such as the maternal serum alphafetoprotein test used during pregnancy.

In addition, says pediatric pulmonologist and medical ethicist Benjamin Wilfond, "doctors typically don't view psychosocial consequences as risks. Surgery is understood to be risky. But risks from genetic testing can have equally serious consequences."[12]

Some of the current problems of inadequate understanding could possibly be alleviated by referring patients to genetic counselors who are knowledgeable about the etiology of genetic disorders and the risks and benefits of testing, and who are less directive than physicians. However, there are only about 2,000 genetic counselors in the United States, and they are disproportionately settled in a few geographic areas (New England, Chicago, and California).[13] And even when counselors are available, physicians vary widely in whether they are willing to refer patients to genetic counselors. A study by Neil Holtzman found that more pediatricians and obstetricians/gynecologists than primary-care physicians, internists, and psychiatrists said they offered some form of genetic counseling.[14]

Quality Assurance in Testing

Even if physicians themselves are knowledgeable about genetics, they may pass along inaccurate information to patients because the tests themselves are faulty, because laboratories run them incorrectly, or because the results are difficult to interpret. There is evidence that it is more difficult to maintain quality in genetic testing than with many other medical tests.[15] If a genetic test for a late-onset disease is negligently done, the erroneous nature of the test results may not be discovered until years later, when the disease eventually manifests. Since, outside of high-risk families, most genetic testing results will be negative, laboratory personnel may be less vigilant.[16] Tests may be performed in specialized laboratories far from the location where the sample is collected, giving rise to more potential junctures at which mixups between samples could occur.[17] False-negative results may occur because the DNA test may not predict all disease-causing mutations.[18] The meaning of the test may depend on analysis of test results in other family members.[19] The test may produce probabilistic data that physicians are not expert in interpreting.[20] Because certain diseases—such as inherited breast cancer—are incompletely penetrant, a positive test result may not necessarily mean that the individual will get the disease.[21] Moreover, the true predictive value of certain tests might never be ascertained if there is a rush to treat based on a positive result on a newly developed test. If, for example, the overwhelming majority of women who have a mutation of their BRCA1 gene undergo radical mastectomies, it will be impossible to study either the penetrance of the breast cancer for women with that mutation or the validity of the test. Randomized clinical trials are needed before women undergo unnecessary surgeries.

Genetic counselors have also raised concerns about the choice of laboratories to which genetic tests are sent. Genetic tests are complex. Hundreds of mutations may exist in a particular gene, yet labs vary in whether they test for ten, twenty, forty, or more of these mutations. Some HMO's insist on using their contract labs for genetic testing, even though those facilities may not be expert in this type of service. If, for example, the contract lab tests for only twenty mutations in the cystic fibrosis gene, the patient who tests negative will think he is not a carrier and may make reproductive decisions based on that information. In actuality, though, he may have a mutation that could have been picked up by a different lab that tested for forty mutations.

In one instance, a patient wanted Huntington's disease testing, but his insurer objected to the high cost of the laboratory the counselor chose. The insurer insisted that the sample go to an in-network lab. Several months passed, with numerous phone calls from the counselor to the lab before she got an oral report of results from the lab. But the lab consultant told her not to talk to the patient about it until she received the written report. The report, which did not arrive until two months later, indicated a negative result. Yet by that time, the patient seemed to be exhibiting signs of Huntington's disease. Although the testing was clearly inadequate, the lab refused to pay for a confirmatory test.[22]

Managed-care organizations and other third-party payers are increasingly limiting access to only those laboratories with which they contract, regardless of members' geographic location or physicians' desires. Unfortunately, the contracting laboratory may not participate at all in genetic test proficiency testing or other quality-control programs, or may perform inadequately under such a program.

Genetic counselors who work at hospitals that have genetic laboratories raise similar concerns. "At our center, it can often take several weeks, even months for a peripheral blood chromosome result to be finalized, whereas the commercial genetics laboratories usually have results in one week," says one counselor. "There is obviously much political pressure to support one's own laboratory, but this is not always in the patient's best interest."[23]

A 1999 study of 245 molecular genetic testing labs found that 36 labs (15 percent) scored lower than 70 percent on a quality-control scale.[24] The researchers conducting the study pointed out the need for improved personnel qualifications and laboratory practice standards.

Nancy Seeger learned about mishaps in genetic testing firsthand. She was told she had a mutation in her BRCA1 breast cancer gene that indicated a 50 percent risk for ovarian cancer. As a result, she underwent surgery for removal of her ovaries.

Then Seeger donated blood to a research study at the University of Illinois and learned that she did not have the mutation. When Oncormed, the biotechnology company that did the original test, was confronted with this information, company officials merely sent her a letter apologizing "for any anxiety or stress the situation may have caused" and refunded the $350 test fee to her.[25]

Although errors have occurred in newborn screening,[26] in karyotyping,[27] in interpretation of linkage studies,[28] in predictive cancer testing,[29] and in Tay-Sachs screening,[30] only a small minority of states have special licensing requirements for laboratories undertaking genetic tests,[31] and those requirements vary widely. New York has a rigorous program for quality assurance,[32] while Maryland merely requires that genetic laboratories perform satisfactorily in any external proficiency testing programs approved by the state secretary of health and mental hygiene.[33]

In fact, some state statutes impede, rather than encourage, quality in genetic services. In some states, traditional malpractice suits have been prohibited with respect to certain genetic services. Seven states have passed statutes prohibiting couples from suing their physicians or laboratories for negligent prenatal genetic testing if their claim is that they would have aborted the fetus if the test had been properly performed.[34] These statutes were the result of lobbying efforts of right-to-life groups.

Federal statutes that could enhance the quality of genetic services do exist, but they are often not complied with. The Food and Drug Administration, under the Medical Devices Act of 1976 and the Safe Medical Device Amendment of 1990, has the authority to scrutinize the components of genetic testing kits that are marketed in interstate commerce. Before such kits can be sold, the FDA requires proof of safety and efficacy in a premarket approval process. The Assessing Genetic Risks Committee of the Institute of Medicine notes that "only a small proportion of genetic tests in widespread use have been reviewed by FDA; these include tests for hypothyroidism, phenylketonuria (PKU) and MSAFP [maternal serum alphafetoprotein]."[35]

Why have so few tests been reviewed by the FDA? Because academic, government, hospital, and commercial laboratories are not marketing tests but are using their own ingredients, known in the trade as "home brews." And the FDA has not aggressively regulated home brews—even though they are being used in tests that are sold to patients who then make crucial decisions, such as whether to have surgery or an abortion, based on the results. In fact, a 1995 survey found that an alarming number of organizations developing or offering "home brew" genetic tests had never contacted the FDA regarding such services. Of the 43 biotech companies and 215 not-for-profit organizations that used "home brew" genetic testing and responded to the survey, only 15.6 percent had contacted the FDA. Furthermore, the FDA has admitted that although it has the authority to regulate these home brew genetic testing services, it is not doing so.[36]

Even when the FDA has assessed a particular genetic test in its premarket approval process, there may be problems in assuring the quality of the test when it is used for other purposes. Maternal serum alphafetoprotein testing was scrutinized for its ability to predict that a fetus was likely to have a neural tube defect. Subsequently the test was also used to predict whether a fetus had Down syndrome, even though the test has neither been specifically approved nor assessed by the FDA for that purpose.[37]

Concern about the poor quality of genetic services is widespread within the genetics industry itself. In a survey of 81 biotechnology companies and 245 nonprofit organizations that offer genetic testing, 67 percent of the biotechnology companies and 75 percent of the nonprofit organizations agreed that "FDA policies, or lack of policies, hinder the development of safe and effective genetic test kits or other products." The vast majority (more than 84 percent) of both types of organizations indicated that there were genetic testing laboratories that lacked adequate quality assurance programs.[38]

The federal Clinical Laboratory Improvement Act (CLIA) has the potential to help improve the quality of genetic tests. This law, which was tightened considerably in the wake of a scandal involving widespread errors in Papanicolaou (Pap) smear tests, covers the hundreds of thousands of laboratories in the United States that provide tests for the diagnosis, prevention, or treatment of diseases or the assessment of health. The law covers only those labs that provide information upon which a health care decision will be made. It does not cover research labs that do not provide identified test results to patients or doctors. CLIA requires labs to participate in proficiency-testing programs (in which blood and other samples are sent to the labs for testing to determine if they can correctly determine the results). CLIA also requires lab inspection every two years.[39]

The problem with CLIA is that labs have to self-identify for CLIA certification in order for the government to know they exist. Few genetic labs have applied for certification.[40] Many researchers at university genetics laboratories that are providing test results to patients and doctors don't even realize they are covered by the law. Moreover, "in many large academic hospitals, the central lab is not even aware of all the labs [in its own institution] that provide services."[41]

In its overall inspection program, the federal government found that 11 percent (2,190) of the physician office labs inspected had serious problems under CLIA, with 80 percent to 84 percent having some kind of problem.[42] Additionally, there is no separate requirement under CLIA for laboratories offering DNA-based genetic testing, so a lab need only satisfy general lab standards under CLIA. Since genetic testing is more complex than much of the other testing offered, there is reason for concern about the quality of information that patients are getting from genetic testing.[43] For example, in a 1995 survey conducted by Neil Holtzman and Stephen Hilgartner regarding the state of genetic testing in the United States, the researchers reported that "several commercial laboratory directors told us they were aware of poor quality laboratories who were offering services."[44] Because of the lack of reliable information available on the quality of standards of labs performing genetic tests, however, providers and patients may never be able to identify these poor-quality labs. In fact, the commercial laboratory directors themselves indicated that they received the information "by word of mouth."[45]

The Task Force on Genetic Testing was created in early 1995[46] by the National Institutes of Health (NIH)–Department of Energy (DOE) Working Group on Ethical, Legal, and Social Implications (ELSI) of Human Genome Research "to review genetic testing in the United States and make recommendations to ensure the development of safe and effective genetic tests."[47] In September 1997 the group issued its final report, which contained its principles and recommendations regarding the appropriate delivery of genetic testing.[48]

The task force strongly recommended that the following criteria be satisfied before a genetic test is offered:

1. The genotypes to be detected by a genetic test must be shown by scientifically valid methods to be associated with the occurrence of a disease. The observations must be independently replicated and subject to peer review.
2. Analytical sensitivity and specificity of a genetic test must be determined before it is made available in clinical practice.
3. Data to establish the clinical validity of genetic tests (clinical sensitivity, specificity, and predictive value) must be collected under investigative protocols. In clinical validation, the study sample must be drawn from a group of subjects representative of the population for whom the test is intended. Formal validation for each intended use of a genetic test is needed.
4. Before a genetic test can be generally accepted in clinical practice, data must be collected to demonstrate the benefits and risks that accrue from both positive and negative results.[49]

The task force also recommended that protocols for the development of new predictive tests be reviewed by an institutional review board (IRB) when the subjects are identifiable; if the results are reported to the patient or the physician, these tests must be conducted in labs certified under the Clinical Laboratory Improvement Amendments.[50] In addition, the task force recommended "strenuous efforts by all IRBs (commercial and academic) to avoid conflicts of interest, or the appearance of conflicts of interest, when reviewing specific protocols for genetic testing."[51] The task force also called for collaboration among federal agencies to share or pool anonymous genetic information, and for extensive follow-up studies regarding genetic tests after they hit the market.[52]

The task force also called for a national advisory group on genetic testing in the Office of the Secretary of Health and Human Services.[53] This committee, said the task force, was necessary "to ensure that (a) the introduction of new genetics tests into clinical use is based on evidence of their analytical and clinical validity, and utility to those tested; (b) all stages of the genetic testing process in clinical laboratories meet quality standards; (c) health providers who offer and order genetic tests have sufficient competence in genetics and genetic testing to protect the well-being of their patients; and (d) there be continued and expanded availability of tests for rare genetic diseases."[54] In June 1999 the Secretary's Advisory Committee on Genetic Testing was established.[55]

Genetic Hubris

People may get misinformation about genetics because their doctors lack knowledge or because laboratory tests are done negligently. But they may also have an erroneous impression of what genetic information signifies, as a result of exaggeration by physicians and biotechnology companies of the predictive capabilities of genetic testing and the therapeutic worth of genetic treatments. Evelyn Fox Keller points out, "Even though, in actuality, genetics remains of quite limited practical relevance to the healing arts, the concept of disease—now extended throughout the domain of human behavior—has increasingly come to be understood by health scientists in terms of genetics."[56] When geneticists and scientists describe "cancer genes," they rarely disclose that fewer than 5 percent of cancer patients have a hereditary form of the disease.[57] For 95 percent of cancer patients, genetic predisposition testing is irrelevant.

Recently, a genetic susceptibility for lead poisoning was purportedly discovered,[58] and the researchers suggested that the finding might lead to genetic screening. Yet Abby Lippman asks, "Do we really want to screen for genes rather than clean out lead to prevent the avoidable damage known to affect the millions of children unnecessarily exposed annually to this toxic agent?"[59]

Geneticist Benno Müller-Hill challenges the hubris with which contemporary geneticists approach the genetic underpinnings of disease. "Let us assume . . . that a gene is isolated which predisposes . . . for schizophrenia. . . . Does this mean that we will understand schizophrenia if we know that it often occurs in people when a certain ion channel or a certain enzyme is damaged? Many scientists would answer in the affirmative, but I would like to say, emphatically, that the answer is no. Understanding a biochemical defect brings us no nearer to the thoughts and actions of the schizophrenic."[60]

In addition, a defect at the genetic level does not necessarily translate to a disease at the level of the person. Even if a genetic mutation is predictive of a disease, it does not indicate when the person will get the disease or how severe the symptoms will be. When the gene for cystic fibrosis was identified, some geneticists advocated screening couples who were planning to have children, and undertaking amniocentesis on the fetuses of carrier/carrier couples, so that the couples could avoid the birth of children who had cystic fibrosis.[61] It was subsequently learned that the disease has a range of severity and that some people with two copies of the mutated gene have no symptoms.

Moreover, learning to diagnose genetic disease does not necessarily mean we will readily be able to treat it. The gene responsible for sickle cell anemia was discovered in 1949,[62] yet there is still no gene therapy for the syndrome. Consequently, the term *prevention* in the genetics context generally means abortion of an affected fetus.[63]

Genetics researcher J. Craig Venter, who has probably sequenced more genes than anyone else in the world, points out how little we know. He published a paper describing the sequence of 40,000 genes, 30,000 of which had never been seen before in science. But sequencing alone tells us nothing about the *functions* of those genes. "Nobody has the slightest idea what they do, what role in biology they play. This opens up tremendous avenues to go forward in science and increase understanding, but I think it puts into perspective what I call our ignorance quotient. Our ignorance is much greater than our knowledge."[64]

Venter and his colleague, Nobel laureate Hamilton Smith, sequenced the 2,000 genes of a particular bacteria. Venter says, "When we were congratulating each other on this completion, I said to him, 'Ham, I'm really delighted you understood all this, because I have to confess, I didn't.' He said, 'Me? I thought you understood this.' And we decided that 2,000 genes is beyond human comprehension in terms of understanding. Humans have 80,000 genes and virtually an infinite number of combinations in these two trillion different cells. So, to have a good understanding of gene function and human physiology is going to take quite a while."[65]

What we *do* know about genes, though, would caution us against moving toward genetic determinism, since the expression of genes is profoundly influenced by a person's environment. "In fact, I'm absolutely certain there is probably not a single condition, physiological or pathological change, that doesn't result in profound changes in cell, in gene expression," says Venter. "Immediately upon putting it in tissue culture and trying to grow it in tissue culture the gene expression pattern changes profoundly.

"It's going to be a long time in many cases, if not most cases, between finding the cause of a genetic disease, the misspelling in the genetic code responsible for Alzheimer's disease or Huntington's disease or cystic fibrosis, and finding a cure or treatment for those diseases. Because these are not trivial experiments to do and effectively getting a correct gene into the correct cells in either single cells or in a whole human is extremely difficult. And it's going to be a very long time before we have effective treatments for all these."

Rather than face the fact that treatment is far off, some scientists make excessive claims about the benefits of gene therapy. W. French Anderson, the lead researcher on gene therapy, predicts a time when doctors will diagnose a patient's genetic disease and, paraphrases one reporter, "give them the proper snippets of molecular thread and send them home."[66]

Since Anderson undertook the first authorized gene therapy on humans in 1990,[67] hundreds of ailing volunteers have been given gene therapy as part of research protocols. Extensive public attention has focused on a study in which children with severe combined immune deficiency (SCID) received gene therapy and improved. However, when the media[68] and scientists (including Bernadine Healy when she was the director of the National Institutes of Health)[69] describe the experiment, they fail to point out that in addition to receiving gene therapy, the children were also receiving standard medical treatments for the disease (administration of enzymes), so it is hard to say exactly what made the difference. Geneticist Michael Conneally points out that the SCID children were getting 50 percent of their enzymes through standard treatments. "You never hear that," he says. "That is never said to you."[70]

National Institutes of Health director Harold Varmus concluded that, even though 567 Americans have undergone gene therapy in approximately a hundred different experiments, "there is still little or no evidence of therapeutic benefit [of gene therapy] in patients, or even animal models."[71] A federally appointed committee investigating gene therapy condemned most of the efforts as "pure hype." Despite the unsuccessful attempts to prove that gene therapy is clinically beneficial, a survey commissioned by the National Center for Genome Resources showed that in 1997, 36 percent of the public who knew gene therapy had been attempted thought it was successful, while only 21 percent knew gene therapy has not been successful in correcting any diseases in humans.[72]

There may also be financial reasons not to pin all our hopes on genetic therapies. University of Michigan genetics pioneer James V. Neel predicts that "individual therapies will be too expensive." As an example, he noted that improved diet and exercise may be more cost-effective at ameliorating adult-onset diabetes than genetic medicine.[73]

There is a crying need for a mechanism to stop the overpromising in genetics and to ensure that people are given adequate information about genetics. Belief in genetics is so strong that some physicians coerce people into learning their genetic status or acting upon it.

Medical Pressure to Use Genetic Services

Genetic services present serious psychological, social, and financial risks. Perhaps because such risks are relatively unique to genetics and infrequently discussed in the medical literature,[74] many physicians view genetic tests as risk-free blood tests and pressure people to undergo them.[75] They may also overestimate the benefits and underestimate the risks of subsequent interventions on people with mutations.

The current push to rush women into breast cancer testing, for example, is evident from the language used by the oncologists and geneticists offering such testing. One oncologist, writing in the *New England Journal of Medicine*, made the startling claim—without documentation—that "the only women at high risk [for breast cancer] who are not being tested are those who are economically disadvantaged or uninformed."[76] She argued that "current guidelines [of limiting testing to women in high-risk families] may be too restrictive, since BRCA1 alterations have been found in patients with cancer without strong family histories." That same physician told a reporter: BRCA1 testing in women between twenty-four and forty "can really make a difference. It is an urgent health care problem to find these women and rescue them."[77] Think about that language—"rescue." Can we really prove that we can benefit the women when we identify them? Does it really "rescue" a woman to tell her she has a particular mutation in her BRCA1 gene if we do not have data on what that mutation means? Does it "rescue" a woman to have a prophylactic mastectomy?

This attitude that breast cancer testing is an unquestionable benefit may lead physicians to offer it inappropriately—without adequate concern for voluntariness, informed consent, and assurances of quality. Already, breast cancer testing has been offered clinically before the predictive value of the test has been adequately ascertained. For example, once it was ascertained that Ashkenazi Jewish women had a higher incidence of a particular mutation, 185delAG,[78] than the general population, clinical testing was offered for the mutation before studies had been completed to ascertain whether those women with the mutation are at higher-than-average risk for breast cancer.[79] When those studies were undertaken, they found that the cancer risk was actually much lower than what patients had initially been told.[80] Some women may not have amputated their breasts had they known the true risk figures.

What researchers and clinicians describe as treatment and prevention approaches based on genetic testing often leave much to be desired. As Francis Collins has pointed out, "Despite the general usefulness of mammograms for the early detection of breast cancer in women over the age of 50, there are no data to instill confidence that regular mammography at a younger age, in concert with self-examination and examination by doctors or nurses, will reduce the risk of death from metastatic breast cancer among very-high-risk women with BRCA1 mutations."[81]

Mark Skolnick, one of the discoverers of the BRCA1 gene, argues that prophylactic surgery can reduce the risk of cancer.[82] For example, a woman with a BRCA1 mutation has a 50 percent lifetime risk of ovarian cancer, and removing a woman's ovaries reduces her risk. But women may not consider that a particular boon, given that the operation will dash their hopes of childbearing by putting them into premature menopause. In addition, some women who have their ovaries removed get the cancer anyway.[83] Even after prophylactic mastectomy, women can get cancer in the remaining tissue.[84]

Skolnick also suggested that women with the BRCA1 mutation might want to have a child at an early age, since early childbearing is correlated with lesser risk.[85] One female scientist has even recommended: "To give our great-granddaughters and their great-granddaughters even better chances of avoiding breast cancer, we should . . . seriously consider the viability of social and health policies to encourage young women to bear children in their late teens and earlier twenties and provide the necessary support for young mothers to complete their education and establish their home and careers."[86]

Such a decision obviously has so many complicated implications for a woman's life that it cannot be equated with a straightforward treatment. And advocating that teens bear children fails to recognize how badly society treats teen mothers.

The pressure on people to use genetic services is particularly strong in the prenatal setting. Genetic counseling is often scheduled immediately before testing, thus not giving the woman a chance to consider all the issues and decide whether she wants the testing.[87] In some clinics, couples see a video describing the risks and benefits of amniocentesis and then immediately undergo the procedure, without having the chance to fully consider the issue.

Sometimes physicians surreptitiously test the blood of pregnant women for carrier status for genetic diseases.[88] In other instances, physicians mislead pregnant women into undergoing genetic testing. In those situations, even so-called voluntary testing may become mandatory. In an innovative anthropological study, Nancy Press and Carol Browner observed physician visits in which physicians offered the maternal serum alphafetoprotein (MSAFP) test to pregnant women. A California regulation requires physicians to offer women the blood test, which measures the level of a fetus's alphafetoprotein that is circulating in a woman's blood.[89] Press and Browner identified factors that caused physicians to exert pressure on women to participate in so-called voluntary MSAFP programs.[90] They found that the physicians did not obtain true informed consent. Physicians offering the testing to women did not reveal the significance of the testing—that it might show that a fetus had spina bifida or anencephaly and that they would be faced with a decision about whether or not to abort. Press and Browner wrote that this kind of faulty interaction between a woman and her health care provider creates a "collective fiction" by "situat[ing] the testing within the domain of routine prenatal care and deny[ing] its central connection to selective abortion and its eugenic implications."[91] Instead, the test was routinely described as "a simple blood test" or as a test to show "how your baby was developing." While testing was supposed to be voluntary, those women who refused testing were hounded by the physician until some consented to it.

The gender of the physician or genetic counselor can also influence the amount of pressure put on women to undergo genetic testing. Women physicians and counselors tend to be less directive, more sensitive to personal autonomy issues, and more concerned with the overall effect that testing may have on the family unit as a whole.[92] The directiveness or nondirectiveness of a physician or genetic counselor may influence not only whether a woman undergoes prenatal diagnosis but also whether she terminates an affected pregnancy.[93]

No data exist on what proportion of American geneticists and genetic counselors direct women to abort affected fetuses, although I have encountered women whose physicians have so pressured them.[94] The type of coercion that women may undergo is illustrated by the case of a woman whose physician discovered prenatally that she would deliver a child with anencephaly. "They [the doctors] said her baby would have more in common with a fish than a human. They said to expect the girl to be as smart as a baboon."[95] International surveys also suggest that pressure from physicians exists. For example, in Portugal, 50 percent of geneticists advocate termination of a pregnancy in which the fetus has cystic fibrosis (as compared to 21 percent in the United Kingdom).[96] This occurs despite the fact that children with cystic fibrosis are of normal intelligence and, with treatment, can, on average, live into their thirties.

Physicians may penalize women for not using genetic services. Health care professionals are more likely to blame women for the birth of children with genetic conditions if the women refuse prenatal genetic testing.[97] This is particularly troubling, since physicians may be less likely to help women who decline testing because "the outcome, giving birth to a child with a condition for which prenatal screening is available, is seen as preventable."[98] Other research shows that physicians try not to spend time with patients when they feel the patients have contributed to the illness.[99]

We are now at a crucial stage with respect to genetic services, as a plethora of such services is being integrated into clinical practice. If this integration is handled inappropriately by physicians, it may mean that patients will be pressured into using diagnostic technologies or treatments that are inappropriate or potentially risky for them. That experience, in turn, may deter patients from seeking genetic services, or even more general medical services, that might actually be necessary.

Impact of Physicians' Beliefs and Personalities on the Provision of Genetic Information

People may also be coerced into acting on genetic information. Medicare told one beneficiary with breast cancer it would pay for her breast cancer gene test only if she would agree to have her ovaries and her other breast removed.[100]

There is no consensus among physicians about whether and when particular genetic tests should be offered. Patients—and even physicians themselves—may not be aware of the extent to which a physician's personality, training, gender, and personal beliefs affect what he or she tells patients about genetic services.

Although there is an overall climate of urging patients into genetic testing, some appropriate candidates for testing may not learn about it. Some physicians might not offer a genetic service because of their own personality traits. A study by Neil Holtzman and his colleagues at Johns Hopkins found that a particular personality trait—tolerance for ambiguity—influenced whether physicians would offer patients genetic testing.[101] The physician's gender, too, can affect what services he or she offers. Surveys of medical students revealed 85 percent of male medical students would not offer artificial insemination to a couple in which the man was at risk for Huntington's disease, despite the fact that half of at-risk men say they are willing to use donor sperm. In contrast, only 17 percent of female medical students would fail to offer artificial insemination.[102]

The physician's ethnic or cultural background may also influence whether, and how, that physician offers genetic services. Many of the genetic tests currently in use search for genetic mutations that indicate illness, predisposition to illness, or susceptibility to illness that affects certain ethnic groups with a much higher frequency. Examples are sickle cell testing for African Americans and testing for Tay-Sachs and the 185delAG mutation in Ashkenazi Jews. These tests are potentially stigmatizing to the affected ethnic group, and there is also cause for concern about whether persons who are not members of the affected ethnic group can appropriately offer these kinds of tests. The Task Force on Genetic Testing wrote that ethnic groups differ regarding their perception of disease and what should be done to avert disease. The task force also found minorities to be seriously underrepresented in the field of genetics.[103] Thus individuals may receive genetic informational counseling that is insensitive or inappropriate in light of their cultural or ethnic beliefs. Given that at least 25 percent of clients who receive genetics services belong to racial and ethnic groups other than Caucasian, this point is particularly important.[104]

A physician's religious beliefs may also influence whether he or she informs patients of the availability of genetic services. When Vanderbilt researchers tried to recruit pediatricians to offer free cystic fibrosis carrier screening to the general population, the pediatricians refused because of their personal beliefs against abortion.[105] They did not want to let couples know such testing was available because carrier/carrier couples might then choose prenatal diagnosis and abortion of an affected fetus. Consequently, people often have to learn about genetic services from sources other than physicians. In a study of 520 women who had undergone amniocentesis, only 36 percent had first learned about the procedure from their obstetrician (while a similar percentage, 36 percent, first learned about the procedure from the media).[106]

Another problem is that, in certain circumstances, clinical decisions are made more for the benefit of health care providers than of the patients. For example, the age of thirty-five for screening through amniocentesis was chosen in Canada in part to ensure that enough women would be available to be tested so that smaller genetics centers could stay in business. If an older age had been picked, there would have been too few women to sustain the volume of procedures needed by the small centers to meet professional credentialing standards.[107]

Informational Materials About Genetics

To help people make decisions about whether or not to undergo genetic services, many policymaking bodies have suggested that patients be provided with written information about the benefits, risks, and alternatives to the proposed services. Yet few studies have been made of the content of the written materials that are provided to people who participate in genetic research, testing, or treatment.

I became concerned about this problem when I saw a brochure for cystic fibrosis testing that was distributed to pregnant women. The brochure asked, "Why should I have this test?" The response was, "To have a healthy baby." A pregnant woman reading the brochure could reasonably have assumed that the test would identify a problem with her fetus that could then be treated. What the physicians writing the brochure meant, though, was that the woman who was tested and whose fetus was found to be affected could abort *that* fetus and then keep getting pregnant and tested until she conceived an unaffected fetus.

When breast cancer genetic testing was first offered, geneticist Neil Holtz-man analyzed the materials provided by the four institutions offering it (two biotech companies, one university clinic, and one for-profit clinic). One of the four dramatically exaggerated the incidence of breast cancer.[108]

Several interesting studies demonstrate deficiencies in the type of information provided to individuals involved in genetics research and clinical services. Robert Weir and Jay Horton collected 103 informed consent forms that were given to participants in genetic research.[109] Only 23 forms explicitly asked for consent before they stored DNA. Of those 23, most did not mention the length of storage, policies for withdrawal of DNA, or whether the individual or third parties would have access to the results of testing of the sample.[110]

Increasingly, DNA is being "banked" for future analysis.[111] In a 1994 survey of DNA diagnostic laboratories, 84 of 93 respondents indicated that they were storing DNA with identifiers. Twenty-nine of these laboratories had no written policies regarding contacting or recontacting the source, or regarding who would have access to DNA and DNA test results. Forty-five laboratories had no type of written agreement with the source.[112]

Mildred Cho, Monica Arruda, and Neil Holtzman analyzed the written materials presented to people undergoing predispositional genetic testing in 169 different programs. They found that the materials "lacked sufficient information about the tests themselves, such as their sensitivity, specificity or predictive value, the purpose of testing, and information concerning patient rights."[113] They found that only 7 percent of the pamphlets for physicians, 26 percent of those intended for patients, and 17 percent of the ones intended for both included any explanation of the risks and benefits of testing. Only 3 percent of those devoted to physicians, 26 percent of those for patients, and 47 percent of those intended for both mentioned confidentiality, voluntariness, or the possibility of discrimination in conjunction with genetic testing.[114]

The type of information included in the materials about genetic testing varied according to whether the materials came from a nonprofit or a for-profit institution. While 35 percent of the nonprofits' materials included information on patient rights, none of the materials from for-profits did. The need for or availability of genetic counseling was discussed in 79 percent of nonprofits' materials, but in only 43 percent of for-profits' materials.[115]

Many genetic testing services are advertised directly to the public, but currently no mechanism exists to review these ads before they are made available to consumers or physicians. Although the FDA must review labeling materials on genetic testing kits as part of the premarket approval, most genetic tests are marketed as services, not kits, and thus are not subject to FDA premarket approval. The only oversight for these informational materials, including ads, is review by the FDA, the Federal Trade Commission, the Consumer Product Safety Commission, or consumer product divisions in states' attorneys' offices *after* complaints about the ads are filed or acted upon. Only after questionable claims or potentially harmful or misleading information is available to the public will regulatory oversight mechanisms kick in. Such a situation is obviously quite troubling, and it led the Task Force on Genetic Testing to discourage advertising or marketing of predictive genetic tests to the public.[116]

Educating people about either genetics in general or the existence of certain genetic services is a complicated matter. At numerous meetings I have attended, geneticists have bemoaned the fact that reporters and the public are not sufficiently knowledgeable about genetics. Virtually every major genetics policy report calls for greater public education about genetics. The Institute of Medicine's report, *Assessing Genetic Risks*, devotes an entire chapter to public education in genetics.[117] Even the National Bioethics Advisory Commission report on cloning asserts, as one of its five recommendations, that the federal government should "provide information and education to the public in the area of genetics."[118]

But what should this information entail? Sometimes geneticists themselves are at fault for the public's misunderstanding of genetics, since they have created a sense of genetic determinism and engaged in overclaiming about genetics. At one meeting I attended, held to educate the press, a prominent genetics researcher told reporters, "We will soon know the genes for every human behavior because we've already identified the three genes that cause hardness in the tomato." University press releases about genetics research make assertions that go beyond the existing data. The university press release about the "obesity gene" emphasized the human applications, even though the research was done on rats,[119] and led to the impression that interventions that worked to slim rats would help humans lose weight. Subsequent research in humans found that this was not the case.[120] Scientists' promising too much about gene therapy has led the public to believe that it is a reality, when actually gene therapy has been tried on at least 567 Americans in more than 100 experiments without any proof of effectiveness.[121] A federal committee investigating gene therapy cautioned: "Overselling of the results of laboratory and clinical studies by investigators and their sponsors—be they academic, federal, or industrial—has led to the mistaken and widespread perception that gene therapy is further developed and more successful than it actually is."[122] Even the educational materials provided to individual patients in connection with a particular genetic service can be misleading. One commercial company exaggerated the incidence of breast cancer in its brochure for a genetic test, perhaps as an honest mistake—or perhaps to recruit more women for testing.[123]

By March 2000, more than four thousand patients had participated in gene therapy experiments. No studies found clear benefits from any type of gene therapy being tested. The only study that remotely indicated a benefit had an extremely small sample size and modest results. In that study, three patients were given a gene transfer to treat hemophilia. They showed a small increase in the amount of clotting factor afterward.[124] Despite the fact that this was hardly a persuasive result, Avigen, the biotech company that funded the research, flooded the media with press releases saying the gene therapy was effective.[125] In contrast, the hemophilia expert who had cloned the gene dismissed the claims as "a lot of hype."[126]

In April 2000, a French study that involved just two patients reported successful results from gene therapy. Researchers treated two children, aged eleven months and eight months who suffered from a rare form of severe combined immunodeficiency (SCID-X1).[127] Though the disease can be cured with bone marrow transplants from a healthy donor, finding an appropriate match can be very difficult. The researchers in France instead removed bone marrow from the two affected children from which stem cells were extracted. Then, using genetically altered viruses, healthy copies of the missing gene were delivered to the stem cells then transfused back into the children.

In many ways, the French chose the disease that is the easiest case for gene therapy. They chose a disorder that is due to a malfunction in a single gene of a person's 80,000 genes. Their approach would not be expected to work for complex disorders that are due to the interaction of several genes (for example, certain diseases such as some cancers and heart diseases).

Also, they chose a disorder where the cells that need the new gene are easily accessible in the bone marrow (and those cells are easy to remove and to insert the appropriate genes into). It is much harder to get genes into tissue such as the brain or liver or lungs. Additionally, they chose a disorder where overproduction of the proteins created by the inserted gene is not harmful. Since researchers currently cannot control how many copies of the gene actually get incorporated in a person's body, there are concerns about too much of an otherwise good thing.

The technology involved is expensive, must be repeated, and is more analogous to something like organ transplantation than to an antibiotic. There are also risks if the scattershot approach used to put genes in the body. They could not work at all if they don't land in the right spot (for example, near a promoter). They could incorporate the genome in a spot that itself causes a genetic disruption, leading to a disease such as cancer. Even if this small study is later replicated, gene therapy pioneer French Anderson cautioned, "Being able to successfully treat SCID doesn't mean you can treat anything else."[128]

Much of the genetic "information" provided in the public sphere is crafted to enhance geneticists' personal and institutional financial standing, rather than to provide accurate information about genetic services. When NIH scientists testify in Congress at budget hearings, they sometimes make excessive claims about the benefits of gene therapy, and about the contribution of NIH to genetic knowledge.[129] It is no wonder, then, that Congress, the president,[130] and the American public[131] are left with a distorted view of what genetics can do.

University researchers, too, are motivated by the potential of commercial gains. Their press releases emphasize "the spectacular promises of research in order to attract corporate funds."[132] Reporters are beginning to refer to this as "press conference science."[133]

"An informed public is the best societal protection from possible abuses of genetic technology and information in the future," notes the Institute of Medicine Committee on Assessing Genetic Risks. "The task, therefore, is to educate the public so that each individual is capable of making an informed decision about seeking or accepting genetic testing and considering personal courses of action."[134]

Yet, even if it is clear that the public should receive certain genetic information, it is important to closely assess the content and manner in which it is presented and to determine who is the appropriate presenter. A Stanford working group points out that "the issues raised by testing are sufficiently complex that educational presentations could easily be consciously or unconsciously biased in any number of ways."[135] The Stanford group also notes that "it is important that educational programs not be provided primarily by groups with financial interests in testing."

Enhancing the Quality of Genetic Services

Problems in the dissemination of information about genetic services and in the quality of those services necessitate caution on the part of patients using the services, action on the part of regulatory bodies to ensure that current laws are followed, and the enactment of new policies and laws to attempt to enhance quality. Professional standards need to be changed so that tests are not adopted prematurely, and so that excessive claims are not made. The need for more in-depth genetic education of doctors and other genetic advisers and more genetic counseling for people who are considering testing is becoming recognized as new uses of genetic services become available. Informational materials provided to the public should be closely scrutinized in order to maintain a neutral and educational presentation. In addition, the economic interests of doctors, hospitals, and companies, and the personal views, beliefs, and backgrounds of physicians should be kept in check so that individuals can have access to the best-quality services possible. Appropriate quality assurance at every step in genetic testing, interpretation, treatment, and counseling must be ensured and the capabilities of these services must be realistically described. Scrupulous efforts must also be made to assure that genetic services are voluntary so that patients are not required to take tests—and suffer inappropriate interventions and discrimination—based on invalid results.

7   The Impact of Genetics on Cultural Value and Social Institutions

      Genetic information has the potential to change the nature of our social fabric by influencing our ideas about individual and social responsibility and by challenging basic societal concepts such as free will and equality. If most people choose to have prenatal diagnosis and abort fetuses with certain disabilities, society may be less willing to provide services to people with those types of disabilities, viewing them as having erroneously slipped through the net of prenatal screening. Already, some physicians and lawyers are claiming that people should have a duty to learn their genetic status and make lifestyle choices—where to live, what type of job to take, what type of insurance to purchase, even whether to bear a child—based on that information. A medical journal article urged parents whose children have a genetic propensity toward skin cancer to quit their jobs and move to a rainy city like Seattle.[1] The journal *Food Technology* predicted that in the future kitchen computers will generate diets individualized to people's genetic profiles.[2] And some physicians and lawyers are saying that it should be considered a crime to give birth to a child with a serious genetic disorder that could have been discovered prenatally.[3]

    Genetics is being appropriated by certain groups as a rationale for social policies. Recently, an organization that seeks to reduce school taxes argued against special education programs on the grounds that since such disabilities are genetic, "responsibility should fall to the medical system, not to the schools."[4] Philanthropic organizations are also beginning to make predictions based on genetics. An article in a philanthropy journal, relying on the book *The Bell Curve*, indicated that some people are genetically predestined to be low achievers, so that it is probably not worth using foundation money to try to enhance their opportunities.[5]

Predicting Future Academic Worth

The availability of genetic testing is changing people's relationship with important social institutions. Schools profess to want genetic information about students in order to help them. The idea is that a student with a gene for dyslexia or, in the future, a gene indicating difficulty with math, might be given additional aid to compensate for the genetic flaw. The problem with such an approach is that it is both too broad and too narrow. Even if such genes were identifiable (and this is a big *if*, given that in recent years reputable researchers from respected institutions like Yale have claimed to have identified genes for complex behaviors, only to have to retract their findings later), having the gene and having the problem are two different matters.

Not all gene mutations are completely penetrant; for many genetic conditions only a minority of the people with the gene mutation actually develop the condition. Often the gene mutation indicates only a predisposition to a problem, and it takes an additional intervention such as a particular environmental exposure to trigger the condition. This means that some children will be labeled as genetically deficient who are not. The implications are profound. The work of Stanford social psychologist Claude Steele on race indicates that students do more poorly if they know they are part of a group that traditionally has not been as strong academically. In studying why blacks do less well on standardized tests than do whites, he discovered a phenomenon known as "stereotype vulnerability." When black and white college students were given a test that was described as simple problem solving, they did equally well. When similar groups of black and white college students were given the *same* test, but told it measured their intellectual potential, blacks performed significantly less well than whites. Steele postulates that blacks have to contend with the stereotype of intellectual inferiority at the moment when they undertake such tests. They try harder, perhaps as a result working too quickly or inefficiently. Steele's research reached similar conclusions with women—and even white males were affected when told that Asians did better on a particular exam. Similarly, when extensive media reports circulated stating that boys did better at math than girls, the scores of girls actually fell off in subsequent years. Widespread genetic testing could create new categories of "losers" and additional types of "stereotype vulnerability," in which people who have gene mutations linked to lesser cognitive ability perform worse than they would have if they had not been told about the mutations.

Teachers' perceptions of their students might be affected as well, and they might give lower grades to children who are performing normally but who have been identified as having an errant gene. This outcome would be consistent with psychological studies in which teachers were told that some students were better than others (when actually there was no difference). The teachers gave the favored students higher grades and more attention, presumably due to the "halo" effect of the positive labels.

In addition to the danger of labeling some children as learning disabled when they are not, a genetic approach to tracking in school can overlook students who actually do need help but whose problems are the result of environmental factors. Schools in Colorado and Georgia have put great emphasis on Fragile X testing to identify children who have borderline mental retardation, but by focusing on that genetic condition—which accounts for only a small proportion of mental retardation—they may be overlooking other needy children. In addition, identification of a child as having the Fragile X condition may cause problems for the child and the family. Some parents have found that once their child's condition is diagnosed as "genetic" through Fragile X testing, schools deny the child opportunities that they give to other students.[6] And insurers have withdrawn coverage for children who are found to have a positive Fragile X test. For one New Hampshire family in which the children had a Fragile X profile, the insurers withdrew coverage for the whole family.[7]

The use of genetic screening in higher education is even more problematic. Some professional schools have rejected applicants who had genes indicating they might die before age sixty. A man who was at 50 percent risk for Huntington's disease was rejected by medical schools that did not want to waste money training someone who might die young.[8]

Predicting Future Health Care Needs

An April 1995 Harris Poll found that 86 percent of people are concerned about the prospect of employers' and insurers' using genetic tests before deciding whether to hire or insure someone.[9] Such a worry is well founded. Recent studies reveal contemporary examples of genetic insurance discrimination.[10] Among people in families with a known genetic condition, 31 percent have been denied health insurance coverage of some service or treatment because of their genetic status, whether they were sick or not.[11]

While individuals themselves might want to know their own genetic makeup to be able to make important life decisions, such information may be used against them by third parties. When Kim Roembach-Ratliff learned through prenatal testing that her newborn child had spina bifida, her insurer wouldn't provide coverage for the child, saying that the disorder should be treated as a preexisting condition. "If we had found out in the delivery room that he had it, he would have been covered," Roembach-Ratliff says. "This genetic information was used against us."[12] Similarly, when a pregnant woman underwent cystic fibrosis testing and her fetus was diagnosed as being affected, her health maintenance organization informed her that it would not pay for the health care costs of the child if she chose not to abort. In that case, though, the decision was reversed after a public outcry.[13]

A woman whose mother had breast cancer was told she could obtain health care coverage but *not* for any treatment of breast cancer.[14] In another instance, a newborn was diagnosed with PKU, a genetic disorder that can cause mental retardation if the infant is not treated soon after birth. She was covered under her father's health insurance, but when he changed jobs when his daughter was eight years old, he was told that she was ineligible for coverage under his new group plan because of her diagnosis, even though her treatment had been completely successful and she was developmentally normal and healthy.[15] Some people who participated in genetics research have lost their health insurance as a result of it, including a man who underwent screening for APC (adenomatous polyposis colon cancer) as part of a research study. Since health insurance companies can exclude people with preexisting disorders, genetic testing provides an enormous loophole whereby numerous diseases can be classified as preexisting because they have their roots in our genes.[16]

Francis Collins, the director of the Human Genome Project, raises concerns that research on breast and colon cancer has been slowed because people have been deterred from participating, fearing genetic discrimination.[17] Fear of discrimination may deter people from undergoing clinical genetic testing, since some insurance companies ask if the applicant has had such testing.[18] In one study of women's interest in genetic testing for breast cancer, 15 percent of women who declined breast cancer testing indicated that they were refusing testing because of worry about losing insurance.[19] Similarly, in a cystic fibrosis carrier study directed by Wayne Grody at UCLA, 11 percent of the subjects experienced concern over what others would think if the test were positive and 30 percent said they would refuse testing if they knew that the results would be available to insurers. Among high-risk individuals offered BRCA1 testing, 72 percent of those with insurance decided to receive their results, while 43 percent of uninsured individuals decided not to.[20]

The chilling irony of genetic testing is that, even in rare cases where a treatment exists, people may be afraid to get tested for the disorder because it might lead their insurer to drop them entirely or cause an employer to refuse to hire them. Such is the case with hemachromatosis, a chronic, fatal disease in which too much iron builds up in the blood. It can easily be treated by periodic withdrawals of blood. Although his father and uncle both had the disease, a graduate student chose not to be genetically tested for it. He was worried about his job prospects.[21] In another case, a man was tested for hemachromatosis and successfully treated. His insurer dropped him nevertheless, on the grounds that he might stop taking the treatment and develop the costly disease.[22]

Relatives of people with Huntington's disease have been refused health insurance.[23] Similarly, "people with von Hippel-Lindau (VHL) disease, a rare hereditary condition that can cause brain and kidney tumors, often find it hard to obtain health insurance because of the expensive surgeries they might need."[24] Parents of children at risk for VHL[25] or for polycystic kidney disease[26] often avoid having their children tested for the mutations because they fear the children will be uninsurable.

Insurance in the United States is based on the ideas of risk spreading and risk sharing. When most people's future health risks are unknown, the future health care costs of a group can be predicted on an aggregate actuarial basis and the costs spread across the whole group. As genetic technologies identify which currently healthy people will later develop which particular diseases, insurance companies have begun charging exorbitant amounts to people predicted to be at genetic risk, or denying them coverage entirely. At first glance, such a policy seems reasonable, akin to charging higher rates to people who smoke. At one company, the job interviewer learned that the applicant's father had Alzheimer's disease and reportedly told her, "I have single parents here, and I don't want their premiums to go up."[27]

As dozens of genes are identified each week, however, the absurdity of this approach becomes apparent. Since each of us has at least five genetic defects, everyone could become uninsurable. Alternatively, if everyone were charged an amount equal to their future medical costs, insurance would entirely lose its risk-spreading benefits.

The genetic revolution is thus calling into question the current structure of health care financing in the United States. This country is one of only two industrialized nations (the other being South Africa) that do not provide their citizens with universal access to health care. Even if the United States is not willing to introduce such an approach, the challenges raised by genetics suggest that careful consideration be given to the possibility of going back to a community rating system, in which the costs of private health care insurance are spread over the entire population of a larger geographic area. Such an approach is already being tried in a handful of states.[28]

Life insurance is viewed differently, as less of an entitlement than health insurance and more of a luxury. Nonetheless, a similar policy question exists regarding whether life insurers should be able to require genetic testing. It seems unfair for someone to be able to learn he or she has the gene for Huntington's disease and then load up on life insurance. But what about the person who has been paying for life insurance for forty years—should he or she face nonrenewal just because a new genetic test reveals previously unknown information?

The policy dilemma can be handled in several ways. Some commentators have recommended that a person be allowed to buy a minimum amount of life insurance—say $10,000—without being subject to genetic testing. This level of coverage would assure that the person received a proper burial. Alternatively, the prohibition on genetic testing could be linked to certain lines of life insurance. The Association of British Insurers has determined that members who offer life insurance will not be asked to undergo genetic tests when life insurance of £100,000 or less is written in the context of a new home mortgage.[29] A third approach, enacted in a few states, allows life insurers to request a genetic test but prohibits them from using it to deny coverage, limit, or affect the terms and conditions of life insurance policies.[30]

Just as insurance discrimination might occur based on genetic information, so might employment discrimination. In the early 1970s, employers discriminated against African American employees and job applicants who were carriers of sickle cell anemia even though carrier status had no relation to the individual's health or ability to perform his or her job.[31] The only significance of sickle cell carrier status was that the person would have a 25 percent chance of having a child with sickle cell anemia if he or she procreated with another carrier.

More recently, a healthy carrier of Gaucher's disease was denied a government job based on his carrier status.[32] Another man was given restricted benefits and denied a promotion and job transfer because he and his son carry the gene for neurofibromatosis.[33] A computer scientist was refused a job when his preemployment physical revealed he had Klinefelter's syndrome, a sex chromosome disorder occurring in one in 450 men.[34] The syndrome can cause sterility but does not affect ability to work.[35] A social worker was fired when her employer learned her mother had died of Huntington's disease.[36] Still another man had a job offer withdrawn based on the claim that he "lied" during a preemployment physical. He had said he was not seriously ill. He had a genetic form of kidney disease, but without any symptoms.[37]

Gail Vance, a molecular geneticist at Indiana University School of Medicine, points out: "If you go to look for a job at age 40 and you have the breast cancer gene, you know that by age 50 half the women with the gene will have cancer. What does that mean to an employer?"[38] A survey of U.S. geneticists revealed that many geneticists would share the patient's genetic information with employers *without the patient's consent.*[39] Physicians are increasingly put in the role of "double agents"—with dual loyalties to the patient and to the patient's school, employer, potential insurer, relative, or child.[40]

According to a 1999 survey of a thousand large and midsize companies, more than half of all new hires are subjected to medical examinations and "dozens of firms don't inform applicants about the types of tests being done."[41] Thirty percent of the companies obtain genetic information on their employees through testing or family histories, and 7 percent use such information in hiring and promotion decisions.[42]

Some people have lost jobs not because of the results of genetic testing but because of their refusal to be tested. Two employees in a Boston telecommunications firm report that they were fired when they refused to provide hair samples for testing. They were concerned that the hair might be subjected to genetic testing.[43]

In September 1995, the San Francisco Legal Aid Office filed a class action suit for the employees of Lawrence Berkeley Laboratories at the University of California, Berkeley, a laboratory funded by the federal Department of Energy.[44] The suit alleged that the lab had tested African American employees for the sickle cell gene without their knowledge or consent, during routine physicals, and had secretly maintained that information in their files. A federal district court sided with the employer, saying that the practice did not invade the employees' privacy, even though they had not been told about the genetic testing, since they had agreed to undergo physical exams and give medical histories. The judge found that given the "overall intrusiveness" of the physical exams and the "large overlap" between the medical histories and the tests, any additional privacy intrusions that resulted from the challenged tests were minimal.[45]

In 1998, however, the Ninth Circuit Court of Appeals disagreed with the district court and held that an employer may not test an employee for "highly sensitive" medical and genetic information without the worker's consent.[46] The court stated that such "illicit" testing,[47] if proved at trial, would be an invasion of privacy in violation of the California constitution and the U.S. Constitution and, because of its differential impact on blacks and women, would amount to job discrimination in violation of Title VII.[48] Judge Stephen Reinhardt, writing for the unanimous three-judge panel, wrote: "One can think of few subject areas more personal and more likely to implicate privacy interests than that of one's health or genetic makeup."[49] Furthermore, Reinhardt added, "it goes without saying that the most basic violation possible involves the performance of unauthorized tests—that is, the nonconsensual retrieval of previously unrevealed medical information that may be unknown even to plaintiffs."[50] The court rejected the argument that the workers had effectively consented to the tests by agreeing to undergo the pre-exam, filling out medical questionnaires, and giving blood and urine samples. The court said none of these acts is the same as authorizing the three tests—sickle cell, pregnancy, and syphilis—at issue.[51] The court allowed the job discrimination claims to proceed, saying the allegations "fall neatly into a Title VII framework."[52] Because women were tested for pregnancy and only African Americans for the sickle cell trait, the court stated that "the employment of women and blacks at Lawrence was conditional, in part, on allegedly unconstitutional invasions of privacy to which white and/or male employees were not subjected."[53] Further, Judge Reinhardt noted that different testing requirements based on sex, race, and pregnancy, even if not unconstitutional, would still be a valid basis for a Title VII discrimination claim.[54]

The Lawrence Berkeley Lab case may presage greater protection for genetic privacy. Courts do seem to increasingly recognize that modern blood tests—such as testing for genetic characteristics or for HIV—are not risk-free. The possibility of stigma and discrimination based on the results is profound. In *Doe v. High-Tech Institute, Inc.*, all students were required by the school to undergo testing for rubella.[55] Before giving a blood sample, one student signed a consent form with the understanding that his blood sample would be tested only for rubella. Nevertheless, an instructor from the school also requested that the student's blood sample be tested for HIV.

The student filed suit against the school, claiming invasion of privacy. A Colorado appellate court reversed the district court's decision to dismiss the claim of intrusion upon seclusion. The appellate court held "that an unauthorized HIV test . . . would be considered by a reasonable person as highly invasive, and therefore, such is sufficient to constitute an unreasonable or offensive intrusion." Furthermore, stated the court, the right infringed here was "the right of an individual to control important health decisions, such as whether to have one's blood tested for a particular disease, condition, or genetic trait."

Statutory Responses to Insurance and Employment Discrimination

Although numerous people have been discriminated against based on their genetic status,[56] few states have legal protections against this practice.[57] Six states have adopted specific statutes prohibiting genetic testing in general without informed consent.[58] But they provide tenuous protection, since there are exceptions to the statutes. In five of the six states, the police can gain access to tissue samples for genetic information.[59] Four of the six statutes have exceptions allowing researchers access to genetic samples without the individual's knowledge or consent.[60]

The laws that have been adopted in recent years to protect against genetic discrimination in insurance also have loopholes. Thirty-four states prohibit denying people health insurance based on certain types of genetic information.[61] Of these, twenty-nine states also prohibit conditioning the provisions of coverage on genetic information,[62] and twenty-seven states prohibit the use of a person's genetic information to set rates.[63] But most of these laws do not protect people from discrimination based on genetic information about their relatives. In twenty-five of the thirty-four states, insurers can easily circumvent the reach of the laws by basing their decisions on family histories.[64] In addition, in most states insurers can get around the laws by discriminating against people not on the basis of their test result but because they previously requested genetic services; only twelve states prohibit the latter type of discrimination.[65] Moreover, because of federal preemption under a statute known as the Employee Retirement Income Security Act (ERISA),[66] the state prohibitions on genetic discrimination do not cover self-funded insurance plans.[67] Yet 85 percent of larger companies (more than five thousand employees) are self-insured.

A 1996 law, the Health Insurance Portability and Accountability Act, provides certain protections so that people cannot lose coverage when they change jobs.[68] Under the act, a group health plan of the new employer cannot deny coverage or apply preexisting-condition exclusions for more than twelve months for any condition diagnosed or treated in the preceding twelve months. In addition, group health plans cannot establish eligibility for enrollment on the basis of health status, medical history, or genetic information.[69] The law is insufficient, however. It does not prohibit genetic discrimination against people seeking insurance under individual plans, such as denial of coverage or exorbitant premiums for coverage. And it does not prohibit group insurers from charging higher rates to a whole group based on genetic information about a particular individual.

In the employment context, sixteen states prohibit conditioning employment on genetic testing,[70] but only a few states prohibit employment discrimination based on family history.[71] On the federal level, the Americans with Disabilities Act (ADA) prohibits employers with fifteen or more employees from refusing to hire or otherwise discriminating against people with disabilities or people who are regarded as having disabilities (unless the disability impedes their ability to do the job in question).[72] In *Peoples v. City of Salina*, a federal district court held it was not discriminatory to stop a person with sickle cell anemia from being a firefighter because the job entails overexertion, stress, low oxygen, and changing temperatures, which can precipitate a sickle cell crisis.[73]

The Americans with Disabilities Act protects people with disabilities who are otherwise qualified to do the job at issue. But it raises profound questions about whether a genetic predisposition without any symptoms should be considered a disability. After much equivocating, the Equal Employment Opportunities Commission (EEOC) in its compliance manual provided guidance about how the Americans with Disabilities Act would apply to an individual who is presymptomatic for a genetic disease.[74] The EEOC wrote that it is illegal for an employer to discriminate against a person based on genetic information relating to illness, disease, or other disorders. As an example, the EEOC indicated that an employer may not refuse to hire an individual just because the person's genetic profile reveals an increased susceptibility to colon cancer.[75] Again, this interpretation may not go far enough; it does not specifically address whether someone can be denied a job because he or she is a carrier of a recessive disorder such as cystic fibrosis and the potential employer does not want to pay the health care costs of potential future affected children. Also, it does not prevent employers from undertaking genetic tests on people, as was done in the Lawrence Berkeley Laboratory case. In that case, the California constitution had a particularly strong privacy provision, allowing the appellate court to condemn the procedure. In other states, employers may legally be able to test employees without consent.

The Lack of Protection for Medical Information in General

The problems of improper disclosure and misuse of genetic information are aggravated by the poor protections given to medical information in general. Secretary of Health and Human Services Donna Shalala says, "Every day, our private health information is being shared with fewer federal safeguards than our video store records."[76]

Patient information is frequently disclosed. Dr. Mark Siegler describes a pulmonary patient who became concerned about the confidentiality of his medical record when he learned that the respiratory therapist had access to it.[77] When the patient threatened to leave the hospital if confidentiality could not be guaranteed, Siegler inquired about the number of health care professionals and hospital personnel who had access to the patient's chart. When he revealed to the patient that seventy-five people had access to the chart in furtherance of his care, the patient stated, "I always believed that medical confidentiality was part of a doctor's code of ethics. Perhaps you should tell me just what you people mean by 'confidentiality'!"[78]

In addition to the revelation of information to enhance patient care, the types of disclosures that a health care provider might be tempted to make range from pure gossip unrelated to the patient's well-being[79] to disclosures to outside third parties that are thought to be in the best interest of the patient, another individual, or society. The individuals seeking disclosure from physicians might include law enforcement officials, public health authorities, relatives of the patient, employers, insurance companies, schools, and lawyers.

It is shocking how little protection exists for private medical information. In some states, laws protect only doctor-patient confidentiality and do not apply to other health care providers. Medical information collected in the workplace rather than the doctor's office is not protected in some states, so genetic screening and monitoring of employees is not covered.[80] Even genetic information collected in a traditional health care setting may be unprotected, depending on *who* collects it. Some existing medical confidentiality statutes protect only medical information in the hands of *doctors* and will not cover genetic information in the hands of Ph.D. geneticists or genetic counselors.[81]

Moreover, even states in which genetic information cannot be released without the individual's permission do not offer sufficient protection. As long as insurers and employers can continue to ask people to "consent" to the release of their genetic information, people will be discriminated against based on their genotype, no matter what the state's law is on medical confidentiality.

Egregious breaches of confidentiality often have no legal remedy. A growing number of employers over the last two decades have chosen to self-insure. In some cases, coworkers and managers have obtained confidential information about employees in that context. When a patient who was HIV-positive was assured by the head of the medical department that the company did not get information about which prescriptions employees filled, he proceeded to obtain his AZT treatment. The pharmacy submitted its mandatory utilization report to the chief administrative officer of the company—listing employees' names and their drugs. The chief administrator reviewed the report with the company's benefit director. The employee's supervisor and other coworkers were told of his HIV status. A jury awarded him $125,000 for invasion of privacy, but the appellate court reversed the decision. The administrator claimed that the drug reports were necessary to look for signs of fraud and drug abuse, although she conceded that such a review could have been undertaken with an initial list that had no patient identifiers. The "important interest" and "legitimate reasons" of the employer, said the appellate court, outweighed what it deemed the "minimal intrusion" into the employee's privacy.[82]

As medical records become increasingly computerized and traded, the potential for inappropriate disclosure grows. "Our systems are now being interfaced and networked in ways never before anticipated," says Kate Burten, a health information security officer at a Boston health care delivery system. "Threats to privacy are greater than ever because data is being captured in electronic form in unprecedented volumes."[83]

Twenty-seven percent of people say their personal medical information has been improperly revealed by insurers, public health agencies, or employers.[84] Some abuses are intentional. A disgruntled Tampa public health worker took a computer disk with a confidential listing of four thousand people who tested positive for HIV and sent it to newspapers.[85] Other disclosures are negligent or inadvertent. When the Harvard Community Health Plan (now Harvard Pilgrim Health Care) computerized its medical records, it erroneously made detailed private psychotherapy notes available to all doctors, nurses, clerks, and assistants.[86]

DNA and Social Decision Making

Protections against inappropriate use of genetics have been slow to develop because so many social institutions view genetic testing as the way to make quick predictions about people's future worth. These institutions have few incentives—or resources—to learn enough about genetics and its application in order to be able to identify the deficiencies in the predictive power of genetic technologies and the professionals who employ them. In other areas of law, courts might be expected to provide a final bastion of protection. But in the genetic era, courts themselves are sometimes seduced by the apparent predictive power of genetic testing.

Courts are increasingly being asked to mandate genetic testing not for medical treatment but to serve as the basis for decisions in lawsuits. Judges have begun to accept—and even require—genetic information in a variety of cases. For judges with a complex, busy caseload, the idea that genetic information may provide legal guidance is seductive. Consequently, the use of genetic testing to answer legal questions is growing, without sufficient thought to the social context or social impact. It is now possible, through DNA testing, to prove that the father of a fifteen-year-old child is a man she has never met. But it may be terribly disruptive to give that man visitation rights just because a biological bond has been proven.

In a South Carolina case, a judge actually ordered a woman to be tested for Huntington's disease (at the instigation of her ex-husband) in order to terminate her parental rights.[87] This precedent may foreshadow genetic battles in all custody cases, creating situations in which divorcing spouses each seek genetic testing of the other in order to find out which one is less likely to get cancer or heart disease and is thus likely to live longer. Such an evidentiary quest may put quantity of time with the child above quality, since the quality of a parental relationship is tougher to measure and prove than the presence or absence of a gene. With this approach, then, the child may not actually end up with the "better" parent. And the genetic predictions themselves may turn out to be wrong.

An even more explosive area of genetic testing could involve personal injury cases.[88] Right now, if an injured individual wins a case involving medical malpractice or an auto accident or other tort, he or she is awarded damages based on statistics about the life expectancy of a person that age. Savvy defendants may begin to require genetic testing of plaintiffs to prove that there is a genetic reason that the plaintiff may die earlier than expected (so that the defendant will have to pay less in damages). When a chemical company was sued on behalf of a child allegedly damaged by the company's toxins, the company persuaded a judge to order genetic testing on the boy in an attempt to prove that his problem was genetic and not due to exposure.[89] Ruth Hubbard and Elijah Wald point out that, in suits against tobacco companies, the companies might be able to avoid liability if they blame the cancers on the plaintiff's genetic "susceptibilities."[90]

Now plaintiffs' lawyers are beginning to order genetic testing on their clients in order to stave off defendants' claims that the person's condition was not due to the defendants' negligence. "As a plaintiff's lawyer, I will often have genetic testing done on a baby to show genetic tests are normal, that there are the proper number of chromosomes and they are the normal size and shape," says Kenneth Chessick, an attorney and surgeon.[91] In one of his cases, a child was born with mental retardation, and the medical malpractice defendant claimed it was a result of Opitz-Frias Syndrome, a genetic disorder. When testing revealed the condition was not genetic, the defendant settled for $1.75 million.[92]

The implications of such legal maneuvers are profound. What if *other* genetic mutations are found when a lawyer orders genetic tests on a litigant? They might be irrelevant to the case, yet have a lasting impact on the individual. The genetic information might cause a person's family, teachers, insurers, and employers to treat him or her differently.

In the coming years, then, anyone who brings a personal injury claim or a custody claim may be forced to undergo genetic testing. Given the enormous psychological and social impact of genetic information, many people who have been injured may be deterred from suing if they think that in the course of the suit a judge might force them to learn their genetic makeup. In the South Carolina suit, the woman absolutely did not want to be tested for Huntington's disease. She faced a painful Sophie's choice and decided to disappear—even though it would mean not seeing her child—rather than be tested.

In addition to the challenges for individuals, the use of genetic information by courts challenges the underpinning of the legal system itself. The entire criminal law system is based on the idea of free will—that criminals "choose" to be bad and thus merit punishment. If offenders are not operating under free will—if, for example, they are insane at the time—they are considered to be not guilty. As more and more geneticists claim they have found genetic markers for antisocial behavior, the legal system will be forced to reconsider the foundation of criminal law. A European research group claims that it has found a gene linked to a propensity to attempt rape.[93] If a man with that gene commits a rape, should he go free on the grounds that he couldn't help it?

Judges seem willing to consider genetic excuses. In two similar California cases, two alcoholic attorneys embezzled money from their clients. The one who claimed that his alcoholism was genetic got a lighter sentence.[94] In another case, a woman who killed her son was found not guilty when her violence was linked to Huntington's disease.[95]

The current view of crime as the result of genetic defects parallels that of a century ago, when it was thought that crime was genetic (and could be avoided by sterilizing people with criminal genes). The popular sentiment fueling it is similar as well. A British reporter described the current trend this way: "Americans, weary with liberal quests for social and economic causes of spiraling crime, are intrigued by the simple notion that some people are born to be bad."[96] Violence is being approached as a medical matter that is financially draining public funds. The Centers for Disease Control and Prevention has declared violence a pressing public health problem.[97] The Los Angeles Times ran an article about "the high cost" of violence, noting that "each year, more than 2 million Americans suffer injuries as a result of violence, and more than 500,000 are treated in emergency rooms." The newspaper estimated that it costs $18 billion annually to care for victims of violence, as compared to $10 billion for victims of AIDS.[98]

Such data have led to attempts to find ways to "cure" people with "criminal" genes through medical means.[99] The type of treatment envisioned might be extremely interventionist. In 1993 a group of scientists identified a genetic mutation that, in a large Dutch family, was associated with males' having borderline mental retardation and abnormal behavior, including impulsive aggression, arson, attempted rape, and exhibitionism.[100] The scientists reported that "isolated complete MAOA [monamine oxidase A] deficiency in this family is associated with a recognizable behavioral phenotype that includes disturbed regulation of impulsive aggression."[101] After the publication of the findings on MAOA (dubbed the "mean" gene by one journalist), talk radio hosts suggested sterilizing people with the gene.[102] In the future, if gene therapy becomes usable, people who are thought to have a genetic predisposition to crime might be subject to such treatment. Harvard microbiologist Jonathan Beckwith points out that "pacification genes may then be added to drug therapy and psychosurgery as tools of social control."[103] George Washington University law professor Harold Green notes that society might also try to restrict marriages when the couple might produce a child with a genetic propensity to antisocial acts, or abortion might be made mandatory if the condition were diagnosed in a fetus.[104]

It is likely that any medical intervention to curtail the manifestation of alleged criminal genes would be applied in a discriminatory fashion. African American individuals are more likely to be prosecuted than white individuals, and African American individuals receive harsher sentences than whites for similar crimes.[105] "Let's just assume we find a genetic link [to violence]," said Ronald Walters, a political scientist at Howard University in Washington, D.C. "The question I have always raised is how will this finding be used? There is a good case, on the basis of history, that it could be used in a racially oppressive way, which is to say you could mount drug programs in inner-city communities based upon this identification of so-called genetic markers."[106]

Genetic Determinism Takes Hold

The trend toward finding genetic explanations for social problems is achieving legitimacy, based on the fact that some of the most influential scientists today are urging a view of humans as genetically determined beings.[107] "We look upon ourselves as having infinite potential," writes Nobel Prize winner and Harvard molecular biologist Walter Gilbert. "To recognize that we are determined, in a certain sense, by a finite collection of information that is knowable will change our view of ourselves. It is the closing of an intellectual frontier, with which we have to come to terms."[108] Harvard zoologist Edward O. Wilson asserts that the human brain is not a *tabula rasa* later filled in by experience but "an exposed negative waiting to be slipped into developer fluid."[109] And James Watson, codiscoverer of the structure of DNA and the first director of the Human Genome Project, has stated, "We used to think our fate was in the stars. Now we know, in large measure, our fate is in our genes."[110]

The social trend today is one of growing genetic determinism, and of making social judgments about people based on their genotype. "Whatever the question is, genetics is the answer," says Baruch College sociologist Barbara Katz Rothman. "Every possible issue of our time—race and racism, addictions, war, cancer, sexuality—all of it has been placed in the genetics frame."[111]

"Geneticists have done a little bit of a disservice because the entire community now feels, these are my genes, therefore I am," says Ed McCabe, a UCLA geneticist and chair of Secretary Donna Shalala's Secretary's Advisory Committee on Genetic Testing. "We need to recognize that a lot of this still has to do with nature and nurture. It's how we live our lives that determines what our risks really are."[112]

New York University sociologist Dorothy Nelkin and University of Pennsylvania historian Susan Lindee point out how the information the public sees about genetics often focuses on explanations of *behavior*. They note that, since 1983, when the category "behavioral genetics" first appeared in the *Reader's Guide to Periodic Literature*, hundreds of articles on that topic have appeared, and "among the traits attributed to heredity have been mental illness, aggression, homosexuality, exhibitionism, dyslexia, addiction, job and educational success, arson, tendency to tease, propensity for risk taking, timidity, social potency, tendency to giggle or to use hurtful words, traditionalism, and zest for life."[113] They also show how these ideas have entered popular culture in novels, movies, soap operas, and advertisements.

During the turn-of-the-century eugenics movement, the scientific beliefs in the predictive power of genetics and the dangers caused by creating individuals with poor genes similarly became woven into the fabric of society. Enlightened political figures such as Theodore Roosevelt[114] and feminist Margaret Sanger[115] believed in the eugenics effort. Genetic beliefs found support on both ends of the political spectrum—among radicals and conservatives,[116] as well as with some religious leaders. Protestant ministers and

Jewish rabbis preached about eugenics and entered a eugenic sermon competition.[117] In 1914 F. Scott Fitzgerald, then a Princeton undergraduate, wrote the song "Love or Eugenics."[118]

Scientists themselves were so convinced of the explanatory power of genetics that in some instances they misrepresented their data to underscore their beliefs.[119] H. H. Goddard, the director of research at the Vineland Training School for Feeble-Minded Girls and Boys in New Jersey, argued that IQ was the chief determinant of moral conduct[120] and that teenagers and adults with mental ages between eight and twelve should be institutionalized and prevented from breeding.[121] He supported his argument by publishing a "study" of a family that he gave the fictionalized name of the Kallikaks, from the Greek word for beauty (*kallos*) and bad (*kakos*).[122] Goddard's book described the descendants of Martin Kallikak. Martin's liaison with a purportedly feebleminded tavern wench supposedly led to descendants who were feebleminded paupers and ne'er-do-wells living in the pine woods,[123] while the same man's descendants with his worthy Quakeress wife purportedly produced descendants who were upstanding citizens.[124] How did Goddard determine the relative traits of the two families? The assistant he sent to interview the branch of the family in the pine woods looked at the strong, healthy father and classified him as sitting helplessly in the corner, saw the children in their scanty clothes and worn-out shoes and found them to have the "unmistakable look of the feebleminded."[125] To persuade readers of the study that these conclusions were accurate, Goddard included in his book doctored photos of the pine woods relatives, with heavy, dark lines inserted around their eyes and mouths to give them an evil, subhuman look.[126]

Today we are once again at a point in history when individual and public interest in genetics is strong—and when fiscal concerns are ruling medicine. As in the past, enthusiasm about genetics has reached even the highest governmental officials. President Clinton, at a November 1996 fundraiser in Houston, described how NIH scientists would make it possible for all infants to be genetically screened and treated shortly after birth. In his 1997 inaugural address, the president said, "Scientists are now decoding the blueprint of human life. Cures for our most feared illnesses seem close at hand."[127] The public, too, is sold on genetics. In one poll, 88 percent of Americans surveyed said they would be willing to give gene therapy to their children to correct a serious disease.[128] Many parents would seek gene therapy to enhance certain traits in otherwise healthy children, including 43 percent who approve of using gene therapy to improve their children's physical characteristics and 42 percent who approve of gene therapy to improve their children's intelligence level.[129]

Today's geneticists reject the earlier genetic assessments as the result of poor science and insist that their own analysis is credible. But the incentives today (obtaining scientific prizes and funding, legitimating the social status quo) are much the same as before, and the results look strikingly similar. Genetics is again being oversold, as an explanation for everything from infidelity to homelessness. As an article in *Science* pointed out, "Today the *Archives of Genetic Psychiatry* is filled with the claims that heredity plays a role in everything from gregariousness and general cognitive ability to alcoholism and manic-depression."[130] Yet the "sighting" of genes for complex traits such as manic-depression, schizophrenia, and alcoholism has often been followed by retractions.[131] For example, the claim that there is a gene for novelty-seeking was dealt a blow when a subsequent study failed to replicate the finding.[132] And, as in the past, contemporary examples of fraud have been discovered in genetic claims. In 1996 researchers at the NIH National Center for Human Genome Research (NCHGR) had to retract substantial portions of five published articles about the genetics of leukemia because a researcher had fudged the data.[133]

Genetics is thought to be the new medical and social panacea. Yet, as Kenneth Schaffner points out, "*purely* genetic explanations do not exist— even for very simple organisms."[134] Richard C. Strohman, professor emeritus of molecular and cell biology, goes further, saying, "The genetic paradigm is dead."[135]

"Learning about the genome for its own sake as a means of understanding biological processes is like learning a language by memorizing a dictionary," says Dr. Claudio Stern, a biologist at Columbia University in New York. "You have all the pieces but you are missing the rules. This does not mean that the dictionary is useless. But when do you come to understand a system? Is it when you understand all the components or understand how the components interact?"[136]

One reason why genes do not give the whole story is that the environment can have a profound effect on the expression of disorders. Before World War II, certain behaviors were thought to be part of the syndrome of schizophrenia, but it was found that they were actually the result of conditions in the large, regimented asylums in which patients with schizophrenia were housed. Changes in institutional policies led to changes in what was considered to be the nature of the disease.[137] Moreover, the prognosis for patients with schizophrenia is better in developing countries than in industrialized countries, which is also suggestive of environmental influences.[138] And studies have found that people with higher levels of schooling are less likely to suffer from Alzheimer's disease after age sixty-five.[139]

Yet the search for genetic solutions has diverted our energies from environmental causes. "Why do we construe childhood poverty as a 'problem too big for ordinary mortals to tackle,' but consider mapping and sequencing all the 50,000 to 100,000 genes we have no big deal? Is children's development disrupted more by genetic loci than by ghetto lead? Do guns or genes alone cause more premature deaths in North America?" asks McGill University epidemiology professor Abby Lippman.[140]

"What if the earlier generation of public health workers had spent their research money on figuring out which arm of which chromosome holds the 'gene for' susceptibility to cholera?" asks Barbara Katz Rothman. "Do we want them to have figured out which people were most susceptible and what was wrong with them that made them so vulnerable? . . . I don't care. I opt for a safe water supply."[141]

The intense focus on genetics may cause researchers to make specious connections and to mistake correlations (or even effects) for causes. In aggressive encounters, a person's hormones, neurotransmitters, and so forth evidence physical changes, some of which may permanently alter a variety of brain and body markers. If in individuals who have been exposed to high levels of violence, researchers view such markers as proof of a genetic cause of violence, they would be confusing effects with causes.[142]

Genetic blinders can also divert researchers' attention from social causes of problems. Biracial children in the United Kingdom are diagnosed as schizophrenic at a much higher rate than either black individuals or white individuals. Those researchers who characterize schizophrenia as a genetic disorder fail to provide an explanation for this epidemiological finding.[143]

Even if accurate genetic predictions were possible, medical interventions might still be inappropriate. Dr. Sarnoff Mednick, a psychologist at the University of Southern California, who published the most commonly cited study linking genetics and crime, argues that even if it could be shown that a gene impaired moral learning by making the autonomic nervous system less responsive, scientists should not try to suppress the genetic trait.[144] "You might also suppress other more positive qualities like creativity," he says.[145]

Certain social institutions are acting as if they have found a crystal ball— in the form of genetic testing—to make decisions about people's worth and claims to social goods. The very notion of responsibility is being redefined in biological terms. Yet this drift toward the geneticization of decisions fails to take into account the psychological, social, and financial impacts of testing people genetically without their consent. It also fails to acknowledge that we are more than just the sum of our genes.

The protections against genetic discrimination are piecemeal. There are real dangers that people who undergo genetic testing or therapeutic techniques will be discriminated against on the basis of their genes. Yet insufficient social pressures exist to change legal policies. Some dramatic ways to exert pressure for protections have been suggested. Congressman David Obey suggested to Francis Collins, head of the Human Genome Project, that perhaps the National Institutes of Health should not give grants to researchers in states that do not protect genetic privacy.[146] Bioethicist Arthur Caplan of the University of Pennsylvania suggested that all genetic testing cease until protective laws are on the books.[147] Both of these recommendations are attempts to bring powerful lobbies—scientists, physicians, biotechnology companies—to the cause of protecting people against genetic discrimination. Currently, it is the people who have been discriminated against who must make the case for such protections in the legislature, and many of them do not want to risk further discrimination and stigmatization by going public. The fight for protections against discrimination in health care has not become a universal fight because many people do not realize that we are all at risk of being defined as part of the asymptomatic ill.

How can we start to control the misuse of genetic information? Three levels of protection are necessary. The first is to ensure that people have control over the genetic information that is generated about them. The second is to give them control of who has access to that information. The third is to prevent discrimination based on genetic information.

8   Which Conceptual Model Best Fits Genetics?

The impacts of genetic technologies are profound. Genetic services can cause psychological, physical, and financial risks to individuals. They can create new forms of discrimination and exacerbate unfair treatment of women, people of color, and people with disabilities. Current mechanisms do not provide sufficient protection for individuals and groups who are offered, undertake, or attempt to refuse genetic services. Appropriate protection would require that extensive attention be paid to voluntariness, informed consent, control of the dissemination of genetic information, quality assurance, and minimizing the potential risks of genetic technologies.

The empirical data on the impacts of genetic testing provide the basis for determining which conceptual model best fits genetic technologies. With a particular model as a starting point, it will be possible to make better policy decisions because those individuals and institutions addressing genetics issues will have to start from a series of assumptions about how the area should be regulated and then justify any departures from that approach. If particular features of a given genetic service make it more appropriate for an alternative model, that alternative will be approved only if completely justified. This procedure will avoid the sort of "extemporaneous" regulation that has characterized much of genetics policy.[1] It will also help avoid the placement of new genetic services into a particular regulatory approach for inappropriate reasons—such as the fact that the approach financially benefits health care providers or helps particular governmental agencies (such as state public health agencies) expand their reach.

The Medical Model

Under the *medical model*, the physician is ultimately the gatekeeper for services and can be quite directive in recommending services. Issues such as quality assurance and confidentiality are generally left to standards set by the medical profession. Medical malpractice suits are seen as the way to enhance the quality of care.

Reliance on physicians to be the gatekeepers for genetic services can be problematic. Their own personal beliefs or personality traits may unduly influence their recommendations. Some doctors refuse to tell women of the existence of prenatal tests because they do not believe in abortion. On the other hand, some physicians are excessively directive—testing people without their knowledge and consent or pressuring women into having abortions. Genetic testing is sometimes undertaken on people—particularly pregnant women—without their advance informed consent.

The information provided to people is sometimes incorrect because of physicians' lack of knowledge about genetics, laboratory error, and the difficulty of interpreting genetic risk information. The medical model for dealing with quality assurance—subsequent malpractice liability—is insufficiently protective, since errors in late-onset testing may not be discovered until decades later. Even with respect to errors that may be discovered in the short term (such as failure to offer prenatal diagnosis or errors in genetic testing), litigation is not sufficient. Some judges state that unless patients tell doctors that they are in a high-risk group for a particular genetic disease, the doctors do not have to inform them about testing for that disease.[2] This puts too much of an onus on patients to be able to decipher genetic meaning from their family history. Moreover, some types of malpractice suits in the genetic realm are prohibited by statute. Wrongful life suits are not allowed by statute in nine states,[3] and wrongful birth suits are not allowed in seven states;[4] in those states, physicians and labs cannot be sued for negligence that results in a couple's carrying to term a pregnancy they would have terminated if they had had all the facts.

The medical model is also problematic when it comes to ensuring that the concerns of disadvantaged individuals—women, people of color, and people with disabilities—are taken into consideration. The physicians who offer genetic services are predominantly white males and have not, in the past, demonstrated sufficient sensitivity to these groups.

The medical model does not seem appropriate for genetic technologies, which can profoundly affect our self-image, intimate relationships, and childbearing plans. It does not provide assurances that the information people receive is accurate or that the services they receive are appropriate. This approach also provides people with information they may not want or are unprepared to use.

The Public Health Model

Although the social policies governing genetics developed initially within a medical model, certain commentators are calling for a *public health model*.[5] In general, the public health approach attempts to prevent disease through education, financing of certain health care services, and in some instances mandating interventions (such as vaccinations). Generally, public health measures have been invoked to prevent imminent, substantial hazards to the population at large through efforts to eradicate infectious disease.

A modest public health approach to genetic technologies might include educating the public about the technologies' benefits and risks and offering them in appropriate circumstances without charge to individuals who cannot otherwise afford them. If a particular genetic service was thought to be exceptionally beneficial, a more aggressive (but extremely controversial) public health approach might require its use.

Some genetics researchers and professionals, as well as some legal commentators, insist that genetic testing and treatment should be governed by a public health model. Other commentators may not realize that they are evoking a public health model when they make policy recommendations or discuss obligations in the genetics context in language associated with public health—for example, by advocating the "prevention" of genetic disease, using analogies to infectious disease, even when the disorders are untreatable and so the prevention is directed toward contraception and abortion.[6] The impropriety of allowing affected children to be born is also implied in articles that point out the financial cost to society of genetic disorders by providing figures on the annual costs of care per patient.[7]

An increasing number of articles advocate a public health approach to genetics, and various commentators urge that people have a duty to learn their genetic status. Dr. B. Meredith Burke states that teens should be required to have genetic tests when they become sexually active. Allowing minors to refuse genetic testing, says Burke, "downplays the moral and legal obligation to protect an innocent bystander"—the fetus.[8] Other policies could be adopted that would further the public health model. One would be the imposition of tort liability for not sharing genetic information with relatives or for not undergoing genetic testing. A California case, *Curlender v. Bio-Science Laboratories*, suggested that a child born with Tay-Sachs, a genetic disease, could bring suit against her parents for not undergoing pre-natal screening and aborting her.[9]

Some commentators go further and suggest that people should be criminally liable for not making use of genetic services. Lawyer and physician Margery Shaw, for example, recommends that states adopt policies to prevent the birth of children with genetic diseases. She suggests that the prevention of genetic disease is so important that couples who decide to give birth to a child with a serious genetic disorder should be found criminally guilty of child abuse.[10]

In many instances, the analysis of why the public health model is appropriate is very superficial. Researchers who wish to use patients' blood samples without their permission to obtain data on the incidence of breast cancer claim that this is appropriate because breast cancer is a "public health threat." However, the mere fact that a disease affects numerous people[11] and is thus a major societal concern does not mean that it is a *public health* problem. Most often, that term applies to imminently dangerous disorders that are highly contagious and thus put the public at large at risk.

Many commentators have advocated a role for genetic testing in public health education. With respect to antismoking campaigns, for example, it is thought that people who knew they had a genetic mutation that predisposed them to lung cancer would be more likely to quit smoking. The idea is that genetic testing takes a general risk and translates it into an individualized risk, thus increasing the person's fear and making him or her more likely to change his or her behavior.

The problem is that genetic information may be so powerful that it makes people too fearful to undertake preventive efforts. That is exactly what happened in a study that incorporated genetic information into a smoking cessation program. People who learned that they were at higher risk for lung cancer were no more likely to quit smoking than people who did not have that genetic information. However, the genetically informed people were more depressed and fearful.[12] The researchers concluded that the use of genetic testing may backfire: "Distress could lead some smokers to deny or to underestimate their smoking problem, which would increase resistance to behavioral change. Distress could also promote smoking to achieve the mood-enhancing effects of nicotine."[13]

In many ways, the traditional public health model does not fit the task of providing education about genetics.[14] The goal of public health educators is to change behavior to prevent disease and to change attitudes and values, which then leads to behavior changes. This is not the proper way to approach genetics.[15] Many genetic disorders cannot be prevented; what is being prevented is the birth of children with these disorders. There is far less societal consensus on the appropriateness of aborting (or not conceiving) affected fetuses than there is on the goals of stopping smoking, preventing heart attacks, or eradicating infectious diseases.

Nor is there a clear justification for the public health approach of *mandating* genetic services. Currently, states use the public health model to mandate the use of genetic services by requiring newborn genetic screening. But there is less need to mandate testing than there was in the late 1960s when newborn screening programs began. If a genetic disorder can be diagnosed in children and prevented or treated during childhood, there is reason to believe in today's litigious climate that private physicians will offer testing—and there is less need for the state to step in.

Moreover, mandatory screening does not meet its public health goal. The government's argument that it has a right to test children, without parental consent, in order to help the children is undercut by the fact that the test results are frequently not acted upon.[16] Most states do not provide funding for the necessary treatments, so poor children do not receive them. In addition, since the public health model does not require that parents be informed of the test and consent to its use, there is a greater possibility of missing cases, since parents cannot check to see if the test has actually been done if they do not know about it. By mandating testing and eliminating the requirement that health care professionals inform patients, society misses the chance to educate people about genetics in general, which would help in other decision-making situations.

The initial purpose of the newborn screening laws was to detect certain diseases early enough for the infant to be treated in a timely fashion. For example, treatment of phenylketonuria shortly after birth can prevent mental retardation. However, some jurisdictions have expanded their mandatory newborn screening programs so that more and more genetic diseases are evaluated from the single blood sample. In some places infants are tested for Duchenne muscular dystrophy,[17] even though early detection will have no influence on the clinical course of the disease. Rather, the information generated is purportedly for the parents' benefit. If the infant has the disorder, that means the mother is a carrier and there is a 50 percent risk that any son she has will be affected. The logic in screening an infant for Duchenne muscular dystrophy is that it will provide information that helps parents make future reproductive plans. But parents could gain such information by testing themselves, rather than by having the government subject their children to testing.

There is some possibility that states may begin mandating genetic testing of adults—in particular, pregnant women. A growing number of pregnant women receive genetic information about their fetuses' well-being through fetal blood sampling, chorionic villi sampling, amniocentesis, maternal serum alphafetoprotein screening, and other technologies.[18] However, the information is obtained at some risk to the fetus itself.[19]

Fetal cell sorting, though, provides information about the fetus without creating a physical risk to the fetus or the pregnant woman. A "simple" blood test is performed on the woman. In the laboratory, complex procedures are utilized to capture minute amounts of fetal blood cells that are circulating in the woman's blood.[20] Prenatal diagnosis is undertaken on those cells, which can determine whether, for example, the fetus has Down syndrome,[21] cystic fibrosis,[22] Tay-Sachs disease,[23] or other disorders.

If states began to mandate such screening, the purpose would be for couples to receive information that would help guide their reproductive decisions. Since, however, the genetic disorders being screened for generally cannot be cured,[24] the only reproductive option in most situations would be to abort an affected fetus. Thus the underlying goal of such an approach would be to encourage the termination of affected fetuses in order to save society the money of caring for such children.

The advent of fetal cell sorting raises an important policy issue regarding women's control of prenatal testing. Because the procedure does not create a *physical* risk to the fetus or the woman, a trend may emerge toward undertaking the testing without the woman's consent. In fact, some of the researchers who developed this technique pointed out that it could be used to screen large populations of women.[25] One group of researchers noted that "because the . . . procedure requires sampling of maternal blood rather than amniotic fluid, it could make widespread screening in younger women feasible. . . . Widespread screening is desirable because the relatively large number of pregnancies in women below 35 years old means that they bear the majority of children with chromosomal abnormalities despite the relatively low risk of such abnormalities in pregnancies in this age group."[26]

Yet mandating prenatal interventions on women would perpetuate the stereotype that women have the primary responsibility for the genetic health of the race. It also would violate a woman's right to privacy, her right to refuse medical interventions, and her reproductive liberty.

Some medical commentators have additionally suggested taking a public health approach to the issue of genetic privacy by allowing physicians to breach a patient's confidentiality and warn the patient's relatives that they may also have a particular gene mutation, or imposing tort liability on people for not sharing genetic information with relatives. They rely on public health precedents that allow doctors to warn third parties about their patients' violent tendencies[27] or infectious diseases.[28]

The case of genetic disease, however, differs from that of violence and infectious disease. In the latter cases, established social policies aimed at preventing violence or the spread of infectious diseases already exist. There are criminal laws against violence and public health laws about reporting infectious diseases and preventing their spread. Thus, breach of confidentiality furthers an established social policy. In contrast, society's position on genetic disease is not so clear-cut. For example, no laws have been adopted to prevent the birth of children with genetic disorders.

The legal cases involving infectious disease and violence could be interpreted as offering no precedent for a privilege or duty for health care providers to breach confidentiality to warn relatives of a genetic risk. In the genetics context, the patient is not *causing* the relative to have the genetic mutation at issue.[29] Warning siblings or cousins about genetic risks will not prevent them from having the genetic mutation (that has already been programmed in at conception), although it might prevent a particularly risky gene/environment interaction. Warning them about their genetic risk may prevent them from conceiving a child with the gene. But such a future risk is not the type of serious, imminent harm that the cases have generally required disclosure about.

Even if the state is viewed as having a valid, compelling interest in furthering the birth of healthy children, mandatory prenatal screening and mandatory disclosure of genetics to relatives arguably do not further that interest. Since treatment for the screened-for disorders is generally not available, the effect of testing is to encourage abortion and to deter carriers from having more children, rather than to promote the birth of healthy children.[30] Because the state would not be able to show that the policy improved the health of potential children, it would likely have to fall back on the argument that such a policy advances a state interest in saving money by discouraging the birth of children with genetic disorders. However, a state interest in saving money should not override a woman's right to refuse a medical intervention.[31] In particular, the potential burden on the state in caring for children has not been viewed as a compelling interest in other contexts.[32]

State-mandated genetic testing for diseases that are not readily treated implicitly devalues people who have disorders for which tests are available. By compelling fetal cell sorting, for example, the government would be influencing the type of children born in our society. This smacks of government-initiated eugenics. Government control of the traits of children is inappropriate, even if—in some policymaker's judgment—some characteristics of the population would arguably be "upgraded."[33]

The empirical data about the potential negative impacts of genetic testing on people's emotional well-being and self-concept, personal relationships, and relationships with insurers and employers would argue against requiring people to find out their genotype against their will. This is especially true in the case of prenatal testing, in which diagnosis of the fetus reveals genetic information about the mother or father or both. In addition to the changes in self-concept that the parent might experience as a result of the revelation of the unwanted genetic information about himself or herself, the parent and the fetus may be stigmatized. If the woman carries to term a fetus who has the gene for Huntington's disease, that resulting child may later be uninsurable and unemployable based on the genetic information collected about him or her without his or her consent before birth. Such a possibility raises a caution against mandatory prenatal screening, at least with respect to untreatable disorders.

Another reason that a public health model seems inappropriate is that there are no standards for what is appropriate testing. Some physicians may seek to obtain genetic information that the woman does not want to know about the fetus or does not want to know about herself. For example, some physicians want to test fetuses for the breast cancer gene even though there is professional disagreement about whether this is appropriate. The lack of consensus about what type of screening should be offered means there is no clear guidance for state policymakers who advocate adopting mandatory screening plans either. State newborn screening programs, which vary in the disorders for which they mandate testing, have already come under criticism. In some instances, states have mandated genetic testing of newborns for certain disorders even when national panels of medical experts recommend against testing for those disorders.[34]

Prevention—the traditional public health goal—does not readily apply to many genetic diseases. Consequently, when the Royal College of Physicians of London listed "prevention" as one of the aims of clinical genetics,[35] geneticist Angus Clarke attacked the idea of making prevention a specific goal:

> If we include such prevention of genetic disorders amongst our aims, we immediately abandon the non-directive nature of genetic counselling in favour of a genetic public health policy, or eugenics. It is impossible to maintain a sincerely non-directive approach to counselling about a genetic disorder whilst simultaneously aiming to prevent that disorder: the opportunities for insider dealing are too great. . . . Its very name clearly conveys the impression that any birth of a child with a genetic disorder represents a medical failure, at least until proved otherwise. This public espousal of prevention, with the unfortunate choice of name, will ensure that the College's initiative is seen as eugenic.[36]

Under the public health model, an effort directed at providing objective information and assuring the voluntary use of the genetic services might be appropriate. Even in the latter situation, however, policy measures are necessary to help assure that participation is truly voluntary, particularly given the pressure that physicians exert on women to participate in so-called voluntary testing programs.[37] To the extent that some physicians pressure women to undergo testing because the physicians want to avoid potential liability for not testing, policies mandating the offering of testing may need to be accompanied by policies exempting physicians from liability if a woman signs a refusal form and subsequently gives birth to a child with a disorder that could have been diagnosed through testing.

The public health approach to genetic services may also raise problems with quality assurance. Some state public health departments have found that as legislatures add more and more tests to the newborn screening mandate, they barely have enough money to test for each disorder, let alone to design programs for quality assurance. Although the Centers for Disease

Control has, at times, undertaken admirable quality assurance programs for state public health departments, at other times the CDC budget has been insufficient to offer such programs. Quality assurance under a public health approach is thus subject to the vagaries of political budget setting and is not a secure way to ensure adequate quality.

The Fundamental Rights Model

The *fundamental rights model* gives greater weight to individuals' decisions about the use of health care services and provides greater assurances of quality. It is applied to health care services that are central to our notions of ourselves. The application of the fundamental rights approach to genetics was initially made in the reproductive context. But the fundamental rights approach fits decisions about other types of genetic services as well. Genetic information is so central to one's own identity that decisions about whether or not to obtain such information and what uses to make of it should be deemed fundamental even if they do not involve reproduction. An additional rationale for protecting genetic services under a fundamental rights model is that it has implications for freedom of association, a constitutionally protected right.[38] Information about a person's genetic status may influence whether someone wants to marry, employ, or otherwise associate with that person.[39]

The fundamental rights model would provide appropriate protections for people using genetic services, in contrast to the medical model and the public health model, which do not. The fundamental rights model would require that participation in genetic services is voluntary and that participants maintain control over their genetic information. Since the medical benefits of genetic testing are in many instances unproven and there are potential psychological and social risks associated with genetic testing, the need to assure that patients make voluntary and informed decisions about whether to participate in testing is particularly significant. It would also require enhanced regulation for quality assurance, since the usual tort incentives for behaving non-negligently are not operating with as great a force in a particular field as they are in other medical areas. There is ample evidence of quality assurance problems in genetics and a need to address them carefully.

Operationalizing the fundamental rights approach would require a careful consideration of the appropriate measures for ensuring voluntariness, adequate information, and quality assurance. Existing legal doctrines would support many of these measures. New laws would be necessary in only a few instances, such as in protections against genetic discrimination.

Ensuring Voluntariness Under the Fundamental Rights Model

The empirical data about the negative impacts of genetic testing on people's emotional well-being and self-concept, personal relationships, and institutional relationships would argue against requiring people to find out their genotype against their will. There is no valid reason to require a person to find out predictive or current disease or characteristic information from his or her genes. In order to assure that people have the right to refuse genetic testing, however, it is necessary to be aware of the context in which genetic testing takes place. Assuring voluntariness requires asking for consent in advance of the procedure and assurance that the individual is not—through situational pressures—making a decision that he or she would not make in the absence of those pressures.

Voluntariness is stressed by the American Society of Human Genetics, the group of doctors who are most knowledgeable about these new technologies. "Careful attention by all parties involved in genetic research should be given to avoiding actions that could be coercive to potential subjects," notes the society.[40] But genetic services have so mushroomed in recent years that most genetic services are not provided by geneticists; they are undertaken by obstetricians, neurologists, psychiatrists, internists, occupational physicians, and others.

Because of the psychological and social risks of genetic testing, the idea that we should not genetically test people without their consent has also been emphasized by various blue ribbon panels of government, ethics organizations, and such entities as the Institute of Medicine. The Committee on Assessing Genetic Risks of the Institute of Medicine recommended: "Voluntariness should be the cornerstone of any genetic testing program. The committee found no justification for a state-sponsored mandatory public health program involving genetic testing of adults, or for unconsented-to genetic testing of patients in the clinical setting."[41] The NIH Workshop on Reproductive Genetic Testing: Impact Upon Women recommended: "Reproductive genetic services should be *meticulously* voluntary."[42] But there are many instances in which this recommendation has been ignored.

Sometimes situational pressures may impede people from making a voluntary decision about participation in genetic services. The providers of such services need to be aware of the psychological needs and preconceived ideas of the person to whom genetic testing is being offered. Such factors as whether or not the individual has a relative with the particular disorder, is married, is highly educated, is a member of a vulnerable social group, is near the age at which a tested-for late-onset disorder is likely to manifest affect one's decisions about whether to undergo genetic testing—and how he or she responds to the results.

To ensure that an individual is not coerced into (or against) testing, the individual's cultural background—including previous experience with and level of trust in the medical profession—needs to be considered. A study by Gail Geller, Neil Holtzman, and others found that "white women thought that breast cancer screening was a way for the medical establishment to make money; African-American women feared possible exploitation if they participated in (any kind of) research."[43] In a further focus group, Jewish women felt a "perceived obligation to do anything they could to advance medical science." Rabbi Elliot N. Dorff, a philosophy professor at the University of Judaism in Los Angeles, indicated that women may have a religious obligation to participate in breast cancer research and testing. Research is "the way we fulfill our obligation as God's partner in the ongoing act of creation. In all cases of illness, no matter what their origin, Jews have an obligation to try to prevent or treat them."[44]

The Johns Hopkins investigators suggest that researchers and clinicians offering testing be sensitive to African American women's distrust of health services. Some doctors assume that breast cancer testing will help African American women come to care earlier, since they now come to care later than white women. However, distrust may be causing them to be late in seeking both care for symptoms and presymptomatic testing.

On the other hand, the investigators advise that those obtaining consent from Jewish women be cognizant of "the 'slippery slope' from perceived social responsibility to coercion."[45] Measures might need to be taken to assure that such women can make autonomous decisions. In particular, potential coercive language should be avoided, such as calling people who reject testing "avoiders."[46]

All individuals should have a right to refuse genetic interventions. Since a child below a certain age does not yet have the capacity to exercise that right, it could best be effectuated by allowing parents the right to refuse testing that has not been clinically proven to have an immediate therapeutic benefit, and by *not* allowing them to affirmatively request such testing for their children. This approach best preserves the ability of the children to make their own decisions about the matter when they reach majority. Children have a right to an open future.[47]

A major source of involuntary genetic testing is secondary testing of tissue samples that the patient provided for other reasons. If a person has had blood analyzed in a hospital or a biopsy done, that tissue is often passed on to researchers and biotech companies to undertake genetic testing on the sample without the person's knowledge or consent. One pathologist tested a woman's tissue for the breast cancer gene without her authorization and then called her and told her she had a mutation in the gene.[48] She thus got unwanted information that could make her and her children uninsurable.

Currently, patients do not realize the extent to which their blood and biopsy samples are used for other purposes. When the tissue is collected in the clinical setting, patients "may expect that the tissue samples will be used only for tests to provide information for their medical care."[49] Even when a person donates tissue specifically for genetic research, he or she is rarely told that it will be stored or that other types of research may be done on the sample.[50] "Many research workers, both clinicians and laboratory scientists, maintain large collections of blood samples from patients with genetic disorders and their relatives," notes geneticist Peter Harper. "So far, they have given little thought to the dangers of this practice as opposed to the advantages. . . . The dangers of misuse of research samples are real and likely to increase as more disease genes are isolated."[51]

Clinical samples from Tay-Sachs carrier screening have been used for prevalence studies of a genetic mutation potentially associated with breast cancer.[52] Commercial products have been developed from patients' blood and other tissue.[53] In fact, according to a survey conducted in the 1980s by a congressional subcommittee, about half of the eighty-one responding medical schools used patients' fluids or tissues for research, accounting for one-fifth of the patent applications that the schools had made in the previous five years.[54] Today that number is probably much higher.

While some people would gladly consent to use of their genetic material for other purposes, some have valid reasons to oppose these additional unconsented-to uses.[55] For some people, such additional uses of their genetic material—immortalizing a cell line, or creating a cloned embryo, or mixing their DNA with DNA from other species—may violate religious beliefs.

Others may have moral, ideological, or political objections to certain uses, such as behavioral genetics research, human gene patenting, or investigations (such as inquiries into race and IQ or gender and mathematics ability) that might disadvantage them as a member of a group.[56] Allowing the use of an individual's genetic material without his or her consent can also create a conflict of interest between an individual and his or her physician, such as when a patient has a particularly valuable genetic material that a doctor wants to sell or patent.[57]

While there might be valid reasons to allow testing of an individual's DNA for identification purposes (such as in the case of forensic testing or paternity testing with proper safeguards),[58] the means of genetic matching chosen should not involve genetic mutations that potentially reveal other information—such as disease predisposition or purported race. Nor should samples collected for one purpose be used for another. Currently, there are few limitations on what additional uses can be made of blood samples collected for forensic identification purposes,[59] or for newborn screening,[60] or for paternity testing. In most states, they could be given to insurers, used by state forensic pathologists to test for genetic mutations supposedly related to violence, or sold to biotechnology companies. Each of these uses may lead to stigmatization of and discrimination against the individual whose sample is involved. Such unconsented-to additional uses should be prohibited.

Since current policies do not generally protect people from being tested genetically without their knowledge or consent, laws should be adopted to criminalize unauthorized genetic testing, except in the case of court-ordered paternity tests or court-ordered tests for law enforcement purposes that meet the criteria of the Fourth Amendment of the U.S. Constitution. Currently, six states have statutes that prohibit genetic testing without consent.[61] But in four of those states inappropriate exceptions are made for research purposes.[62] Such exceptions conflict with ethical and legal principles regarding research in the United States, which hold that research is not a matter of conscription. People may not be made guinea pigs without their consent. Consequently, people's genetic material should be used only with their express consent.

Mechanisms may need to be developed to assure that people do not feel coerced into genetic testing. People must be allowed to exercise what feminist and disability activist Anne Finger calls "the right to choose not to choose." We must scrutinize the sources of information and the information that people get in the context of genetic testing, including what they are told about the level of disability of the disorder for which testing is undertaken.[63]

Information Provided to Individuals Under the Fundamental Rights Model

The principle of voluntary testing cannot be achieved unless people are given adequate information in advance of being asked to consent. The provision of adequate information presupposes an informed and knowledgeable health care provider base, but unfortunately no such foundation is currently in place.[64] A societal mechanism needs to be developed to train physicians, nurses, and other health care providers or lay information providers about genetics—and about the limits of genetic testing and treatment technologies. These information sources should be helped to recognize their own potential biases, such as the fact that male physicians are generally more directive than female physicians and that primary care providers are more directive than geneticists. They should also be made aware of how failure to provide adequate information and counseling to individuals in the context of a particular genetic test can diminish the individual's trust for the health care provider and make it less likely that the individual being tested will be willing to use other genetic services.[65]

Under the fundamental rights approach, individuals would be entitled to extensive information about potential genetic services in the nonreproductive context (such as information about the availability of testing for genes associated with cancer), just as they are entitled to such information in the reproductive context. The responsibility to provide this information to their patients would give physicians an incentive to learn about genetics and genetic services. Such an impetus could combat the current widespread deficiency in physicians' knowledge of genetics.

The information that should be disclosed when genetic services are offered includes those facts that might be *material*—that is, relevant—to the individual's decision about whether or not to undergo testing. The person considering genetic testing should be told of the potential benefits of testing as well as the potential psychological, social, and financial risks that can result from the genetic information, and its implications for the individual, his or her partner, a resulting child, and other family members. The person to be tested should be told whether he or she is at a particular psychological or social risk—perhaps because of his or her form of employment, insurance, family structure, or past psychiatric problems. The person should also be told about the practices of the health care provider or institution regarding who will have access to results. All of this information should be conveyed *before* testing is undertaken.

The importance of informed consent is not recognized by all health care providers. In a study assessing physicians' attitudes toward informed consent for breast cancer susceptibility testing, 18 percent underestimated the importance of patients' informed consent and 34 percent underestimated the need to discuss possible insurance discrimination.[66]

Health care providers may be concerned about the ability of patients to understand genetic information. Although there is some evidence that those without previous experience with genetic disease do not understand the meaning of genetic tests, education through a brochure, video, health care provider, or some combination of those sources can ensure that more individuals have an adequate understanding of the nature of a particular genetic disease, its pattern of inheritance, and the meaning of test results.[67] The existing legal doctrine of informed consent would seem to require such an effort. There are also good policy reasons to encourage the medical community to make efforts to ensure that patients are informed and do understand. The massive efforts to map and sequence the human genome (financed with over $3 billion of taxpayer money) will lead to an increasing array of diagnostic genetic tests. People will need to understand genetic information in order to decide whether or not they want to undergo such tests in order to make medical and personal decisions.

People should be told of the potential significance of the information that they will receive through genetic testing. This means that it may be improper to report genetic mutations whose meaning is unknown. Right now there are more than a hundred mutations in one of the breast cancer genes. For most of them, we do not know if they mean a woman is at a higher risk of breast cancer or not. A biotechnology company, Myriad Genetics, reports any mutations in a patient's breast cancer back to the physician's patient, even if no studies have shown that that mutation leads to cancer. But what if that woman has a prophylactic mastectomy that turns out later to have been unnecessary because that particular mutation has no relation to increased risk? There has been litigation by women whose breast cancer biopsies have been misinterpreted, leading to unnecessary surgery and radiation therapy.[68] There could be similar litigation based on inappropriate genetic information.

The information that is given in advance of genetic testing should describe the potential benefits of the testing, its limitations, and its potential impact on the person's life. Caryn Lerman and Robert Croyle point out that people should be advised in advance of testing of "the limitations of genetic testing and the lack of conclusive evidence concerning available options for prevention and early detection. If genetic testing commences without previous awareness of these limitations participants are likely to be disappointed with the results and may be more vulnerable to the adverse psychological consequences of testing."[69]

This information is in keeping with what people appear to want to know about genetic testing. The reasons that women give for not undergoing breast cancer genetic testing, for example, reveal the type of information that should be provided in advance of testing. A focus group study by Gail Geller, Neil Holtzman, and their colleagues at Johns Hopkins University found that women "were very interested in BRCA1 testing until the limitations and uncertainties associated with the test were understood. In particular, when they learned that most breast cancer is not associated with a BRCA1 mutation, that effective means of preventing breast cancer have not been proven, and that there are risks associated with disclosing test results to employers and insurers, participants [including women from high-risk families] began to question the value of testing."[70]

Eighty-eight percent of women seeking BRCA1 testing in one study did so for reasons of prevention. However, "fewer than one-half of these women were aware that having one's breast or ovaries removed will not definitely prevent cancer."[71] People often seek genetic testing to end uncertainty,[72] but they may mean something different by that than the clinicians do. A genetic test can tell you whether you have a particular BRCA1 mutation, but it cannot tell that you will or—especially, that you will not—get breast cancer.[73] Thus, women may be less interested in BRCA1 or BRCA2 testing when they learn that, despite its being touted as "tests for breast cancer," the testing actually can predict only an increased probability of a small percentage of such cancers.

People undergoing genetic testing should be told what future decisions they will face based on the results of that testing. Among them might be decisions about additional testing (such as amniocentesis after a positive maternal serum alphafetoprotein test) or consideration of whether to carry the pregnancy to term. It is troubling that some doctors describe maternal serum alphafetoprotein testing (an initial prenatal test for spina bifida and anencephaly) as a "test to see how your baby is developing" without informing women that, if that test and confirmatory tests are positive, there is no effective prenatal treatment and thus the woman will be faced with a decision about whether to terminate the pregnancy.

In its "Statement on Storage and Use of Genetic Materials," the American College of Medical Genetics (ACMG) notes that "in the case of most genetic tests, the patient or subject should be informed that the test might yield information regarding a carrier or disease state that requires difficult choices regarding their current or future health, insurance coverage, career, marriage, or reproductive options. The objective of informed consent is to preserve the individual's right to decide whether to have a genetic test. This right includes the right of refusal should the individual decide the potential harm (stigmatization or undesired choices) outweighs the potential benefits."[74]

It is also important for people to know the impact and range of manifestation of the disability for which testing is being done. The individual for whom testing is proposed should also be given detailed information about the specific biological or psychological limitations associated with particular disabilities and what they mean in terms of day-to-day functioning, the services that are available to benefit children with specific disabilities in a particular area, and potential contact with a person with the diagnosed disability, as well as with a representative of a disability rights group and an independent living center.[75]

Under a fundamental rights model, people would also have control over their personal genetic information. Such an approach is already evinced in certain states that prohibit the use of genetic information in employment decisions or insurance decisions,[76] as well as in medical organization guidelines that prohibit physicians from routinely giving genetic information to employers.[77]

Quality Assurance Under the Fundamental Rights Model

The fundamental rights model would also put greater emphasis on quality assurance than the other models do. Currently, there are insufficient safeguards for people who are undergoing genetic testing. Tests are being adopted prematurely, and errors are occurring in analyses and interpretations.[78]

A variety of policies could be adopted to enhance quality in genetic testing. A lengthy research period could be required for each genetic test before it is applied clinically. Individuals and/or institutions that perform testing could be required to meet certain standards. More stringent proficiency testing could be required for genetic tests.

One step toward assuring the quality of genetic tests would be to develop stricter guidelines about the levels of sensitivity and specificity that tests must meet before they are introduced clinically. As a related quality assurance measure, the report of a Stanford Working Group on genetic breast cancer testing suggested that "the federal government should regulate the introduction and use of genetic tests for conditions, disease susceptibility, or carrier status."[79] Currently, if DNA analysis is offered as a test kit, to be sold, it is considered a medical device to be regulated by the Food and Drug Administration (FDA). If, however, it is offered by laboratories using their own reagents ("home brews"), it is not currently considered a medical device.[80] The FDA is proposing to expand coverage, with a system that would work as follows: FDA investigational approval would be needed for human-subjects research involving genetic tests. During this period of preliminary approval, health care providers administering the tests would be required to collect additional data about the safety and efficacy of the test.[81] Only if the test was shown to be medically, psychologically, and socially safe—*as used*—would final approval be granted.[82]

To further help ensure quality, *all* testing programs should collect confidential data to address the unanswered questions about "the risk conferred by the mutations, the nature of the cancers they generate, [and] the consequences of testing on those who receive it."[83] This procedure would appropriately treat new genetic tests as still being in a research phase.

Another method of enhancing quality is to set standards for the practitioners and institutions that provide the services, including a requirement that genetic testing be done only by physicians or genetic counselors who have received specialized, additional training related to the particular disorder being tested for.[84]

Proficiency testing is also central to improving the quality of genetic services. Unfortunately, many of the existing proficiency testing programs are voluntary, and some of the most problematic laboratories may not be participating. In addition, most proficiency programs have no mechanism for penalizing laboratories that repeatedly fail the test, or even to warn consumers about which laboratories provide shoddy services.

Because of these drawbacks to current proficiency testing, participation in quality assurance programs should be mandatory. There is also a need for external blind proficiency testing, in which samples are submitted to a laboratory under a fictitious patient's name.[85] In many instances, the recipient lab knows that the sample they receive is for a proficiency test, and thus laboratory personnel may take extraordinary care with that sample; hence, the results on the tests may not accurately reflect the general quality of the lab. In addition, to the extent that voluntary proficiency testing is continued, the testing organizations should disseminate materials listing the labs that passed.[86]

An additional way to enhance quality is through financial incentives. The Task Force on Genetic Testing recommends that health care payers limit reimbursement for genetic tests to only those labs on published lists that show the lab has satisfactory performance.[87]

The traditional mechanisms that we have relied on in the past to protect us from poor-quality medical services are not working here. Previously "neutral" scientists are now commercially driven, and clinicians' and hospitals' review boards are not sufficiently aware of the psychosocial and social risks of genetic services. "As approval from an ethical review committee has to be obtained before a research project can proceed, it might be thought that this process would ensure that suitable consent was indeed obtained," writes geneticist Peter Harper. "Members of ethical committees, however, are often not experts in genetics and are likely to be more concerned with the dangers and discomfort of procedures or treatments than with the consequences of genetic information. Thus, a project entailing a simple venipuncture [blood draw] may pass review easily in comparison to an invasive measure without full consideration being given to the serious potential effects of detection of a genetic defect."[88]

Legal Justification for the Fundamental Rights Model

As study after study has shown, the impact of genetic services is profound. The medical model and the public health model would, in a de facto or de jure manner, require people to undergo certain forms of genetic testing. In contrast, the fundamental rights model would allow only voluntary testing. In addition, only the fundamental rights model would be concerned with the impact of third parties' use of an individual's genetic information and would provide adequate assurance of quality. Consequently, it seems that the fundamental rights model—protecting voluntariness in testing, control over the uses of one's own genetic information, and accuracy of test results— most closely comports with people's needs.

The studies on the impact of genetic information justify a fundamental rights approach that requires that people be given appropriate information about genetic services and allows them to refuse genetic services. Such a model is not hypothetical. Numerous legal precedents exist that could be used currently to assure people the protection of the fundamental rights approach. The common law right of bodily integrity[89] and the right to refuse medical interventions[90] could be used as the basis for a suit based on unauthorized genetic testing or treatment. If the testing or treatment is part of a state or federal program (for example, if it is established by a statute or regulation or if a state or federal university or clinic or agency undertakes such services without consent), additional constitutional protections come into play to protect the right to refuse the intervention: privacy protections for certain personal information; protections against unreasonable searches and seizures; protections of bodily integrity; and protections of reproductive decision making and decisions regarding childrearing.[91]

Medical information is protected as private, in part because of the psychological, social, and financial risks of its disclosure.[92] Common law privacy protections exist for certain types of medical information,[93] as do federal constitutional protections.[94] Some mandatory genetic testing would provide medical information about the woman or fetus to third parties (the laboratory personnel, the woman's physician), which could arguably be a breach of privacy.[95] Such testing would violate the privacy right not to know medical information about oneself.[96] The right of informed consent also includes a right to refuse medical information that is offered by physicians.[97]

An individual could assert a Fourth Amendment right to refuse the collection of blood or other tissue for a genetic test that was mandated by law or that was undertaken by a government institution, such as a state university medical school. Mandatory blood testing is considered a search and seizure that must comply with Fourth Amendment standards that balance the nature and quality of the intrusion against the strength of the given state interest.[98] Under such an analysis, mandatory testing of an arrested individual's blood for HIV infection has been found unconstitutional under the Fourth Amendment.[99] Similarly, mandatory HIV testing of state employees working with developmentally disabled clients was enjoined as an unreasonable search and seizure under the Fourth Amendment, since the employees' privacy interests outweighed the state's interest in preventing the low risk of clients contracting AIDS from employees.[100]

The right to refuse genetic testing is also protected by common law[101] (and in some cases, constitutional law)[102] protections of an individual's bodily integrity, as well as by constitutional protections of reproductive autonomy.[103] Recent cases have begun to recognize women's right to refuse invasive interventions, such as cesarean section, during pregnancy;[104] it could be argued that such protections would also extend to unconsented-to genetic testing, such as fetal cell sorting to determine the genetic status of the fetus.[105]

Not only is an individual's body private territory, protected by constitutional law and tort law, but the information generated through the use of genetic technologies is private as well. Yet, despite all the precedents protective of individual medical decision making, proponents of mandatory genetic testing (such as mandatory fetal cell sorting on pregnant women) might argue that the process creates only a minimal burden on an individual and thus should not be viewed as an infringement of a person's constitutional rights. The view of a blood test as creating minimal risk is present in some Fourth Amendment cases.[106] Moreover, the cases holding that pregnant women have a right to refuse cesarean section turned, in part, on the fact that such operations are massively physically invasive. In *In re A.C.*, for example, the court held that it was improper to order a cesarean section on an unconsenting woman, but stated: "Our discussion of the circumstance, if any, in which the patient's wishes may be overridden presupposes a major bodily invasion. We express no opinion with regard to the circumstances, if any, in which lesser invasions might be permitted."[107]

Even though some courts have viewed blood tests as insignificant, however, there is reason to believe that genetic tests are different. The federal government, for example, treats them as different. While certain other blood tests used in federally funded research may be exempt from full institutional review board scrutiny since they are viewed as entailing "minimal risks,"[108] the federal Office of Protection from Research Risks has indicated that genetic tests present greater than minimal risks due to the psychological risks and social risks, including "stigmatization, discrimination, labelling, and potential loss of or difficulty in obtaining employment or insurance."[109]

Other precedents exist that could be applied to implement a right to control secondary uses of one's genetic material, such as when researchers want to undertake subsequent testing on a previously collected sample. Here, the individual's bodily integrity is not violated, since the sample exists outside of his or her physical persona. But, if the patient does not explicitly have control over subsequent uses, the physician may have an incentive to remove more tissue than is necessary for the individual's own health care purposes. In *Moore v. Regents of the University of California*,[110] where a doctor allegedly patented a patient's cell line without the patient's knowledge or consent, the California Supreme Court held that the law requires that physicians disclose *in advance of removing tissue* whether there is a chance that they will pursue scientific or commercial research on it.

Giving people the right to refuse genetic testing or research on their tissue samples is in keeping with a vast body of legal decisions. Cases dealing with informed consent,[111] fiduciary duties,[112] the disposal of body parts,[113] tissue transplantation,[114] and relatives' rights to make decisions about a deceased person's organs and tissues[115] all create constraints on what researchers may do with tissue and what information they owe subjects and relatives. Taken together, these precedents indicate that patients are entitled to certain information before tissue is removed. They also underscore the fact that people's psychological welfare can be enhanced by giving them (or their relatives) control over what happens to their bodily tissue after it is removed from them or after they die.

Although the fundamental rights model would assure people's right to refuse genetic services, it would also protect those people who decide to undergo such services. Quality assurance would be enhanced. And there would be equal protection under the law, whereby a person's genetic status would not be a permissible basis for discrimination against him or her by societal institutions. This would remedy the fact that the genetic discrimination is now permissible in many states.

Drawing the Line in the Genetics Realm

At a meeting on human cloning, scientist W. French Anderson said, "The cloning issue is, of course, a significant one in its own right, but it really fits into a broader picture of the genome, our genes, what makes us who and what we are."[116]

New genetic services will be proposed in the months and years to come. Society may make a decision, as California, Louisiana, Michigan, and Rhode Island did with their bans on creating children through cloning, that certain tests or technologies should not be offered. Around the world, a variety of commentators are beginning to raise questions about the types of genetic services that should be banned. British geneticist Angus Clarke is frustrated that "there has been a tendency for molecular genetic and other fetal diagnostic tests to be adopted as a matter of course once they become technically feasible, without a careful assessment of the ethical issues involved. Our justification for this has been the claim that the ethical questions are faced, and answered, by the families who consult us: it is their decision and we wash our hands of any responsibility."[117]

In the clinical sphere, tests are being offered for adults before their clinical significance is known, again allowing doctors to evade responsibility for the services' disruptive and confusing impacts by claiming that it was "the patient's choice." Bernadine Healy, former director of the National Institutes of Health, has said in the breast cancer context, "Physicians and their patients must be wary of overestimating the benefit to the patient or her family if the right information is applied in the wrong way or applied too soon." Healy points out that commercial incentives push testing into the clinical setting before scientific studies can be done to learn what the test results actually mean.[118]

With prenatal testing, the slope has been even more slippery. A wide range of testing options has been offered without sufficient thought to the impact on the potential parents or on existing individuals with those genetic mutations. Such an approach ignores the fact that couples are losing wanted pregnancies. Some couples abort fetuses with an XO chromosomal compliment, known as Turner's syndrome, which is not a cause of mental retardation and rarely causes serious medical problems.[119] "Even if they do understand the medical prognosis, many couples find it tremendously difficult to continue with an 'abnormal' pregnancy, and suffer whatever the decision they might take," says Angus Clarke. Yet the emotional anguish, the subsequent guilt and depression, "are all too easily and frequently ignored" by the genetic service providers.[120]

An increasing number of dizzying "choices" will be made available to potential parents in the coming years. Twenty-six percent of genetic counselors say that a test for tendency toward violence should be developed. Fourteen percent support prenatal testing and/or abortion for tendency toward violence, IQ, body shape, height, fertility, and sexual orientation.[121]

Clarke and others are suggesting that some forms of prenatal genetic testing should not be offered. "Examples might include genes that predispose to a particular disease later in life (such as Alzheimer's disease, Parkinson's disease, or breast cancer)," write law professor George Annas and geneticist Sherman Elias. "From the perspective of the fetus—life with the possibility—or even the higher probability—of developing these diseases in late adulthood is much to be preferred to no life at all."[122] But the same might be said of a child with Down syndrome, which is routinely tested for prenatally. So where do we begin to draw lines?

Angus Clarke points out that although we allow abortion by choice, we generally override parental choice and do not allow abortion for the purpose of sex selection. "Because termination of pregnancy on the grounds of fetal sex when the reason is social rather than medical would be tantamount to a declaration that females are of much less social value than males, society is not willing to make such a statement, which would have profound implications for how women are viewed in society, and also for how women view themselves."[123]

Clarke says, "We draw some moral lines for social but none for genetic termination of pregnancy." He suggests that "to leave all decisions to the discretion of the parents indicates the low value that our society places upon those with genetic disorders and handicap."[124]

The need to draw a line is underscored as we move into an era where genetic enhancement might be possible. In a national public opinion poll, 36 percent of respondents said that controlling their potential child's sexual orientation through genetic technology would be very important to them.[125]

Several strategies have been suggested for drawing the line. One is to focus on the economic costs of caring for the resulting child. This criterion seems to be a particularly cruel and crude way of handling a profoundly important analysis. Looking at near-time or even long-term estimates of medical costs of a particular individual is deceptive. Is the "cost" of a healthy child with Down syndrome actually more than the "cost" to society of educating certain individuals with normal health and intelligence who may be perpetual graduate students, men who hold responsible jobs but commit domestic violence, or people who smoke and need treatment for cancer? And if we did decide to favor potential children based on economic grounds, wouldn't that cause us to reconsider the issue of aborting female fetuses? Gender-selection abortion may be done for economic reasons since, in many cultures, males are "worth" more. Even in the United States, women make only seventy-five cents for every dollar men make.[126]

A second approach would be to consider the likely degree of medical or physical disability caused by the condition at issue and the severity of suffering, the input of people who themselves have the genetic condition at issue, and, to a lesser extent, the input of health care providers who treat such individuals. This approach needs to be supplemented with a broad analysis of the type of society we want to construct. What kind of physicians do we want? What are we willing to require of people in planning their personal and social lives? What message do we want to send about the worth of people with disabilities? What psychological and social inputs should people have to endure in the name of genetic "choice"? How much hype are we willing to believe before we begin to question the motivations and the data of the researchers? How can we begin to counteract the growing genetic determinism?

Combating Genetic Hubris

As historian Daniel Kevles and genetics researcher Leroy Hood have emphasized, "In its ongoing fascination with questions of behavior, human genetics will undoubtedly yield information that may be wrong, or socially volatile, or, if the history of eugenics is a guide, both."[127] As we attempt to develop legal policies to deal with genetics, we should learn not only from our history but also from what I will call our future history: How will later generations judge our actions?

The fundamental rights approach assures that genetic testing would take place only voluntarily, that extensive information would be provided in advance, and that quality assurance mechanisms would be in place. Such an approach could generally be implemented through professional guidelines, individual clinical practice protocols, or legislation. Protections against discrimination, however, would have to be implemented through federal or state legislation.

The fundamental rights model gives greater weight to individuals' decisions about the use of health care services, provides for enhanced information to individuals in advance of using services, protects individuals' ability to refuse services, provides greater assurances of quality, and gives individuals greater control of the information generated about them. It requires greater justification for governmental restrictions on genetic services. In some circumstances, it also requires public funding for services for people who cannot afford them.[128]

The policy model we choose will be the caretaker of our values as we decide upon the proper uses of genetic technologies. It can play an important role in assuring we do not repeat our mistakes. At a minimum, the law should guarantee that individuals can refuse genetic testing and other genetic interventions, that they receive accurate information upon which to base their decisions about using genetic technologies, that they control access to their genetic test results, and that they are protected from discrimination based on their genotype.

Today genetic testing is being undertaken for a wide range of conditions—carrier status, genetic anomalies in fetuses, and late-onset disorders such as breast cancer—without sufficient attention to its psychological impact nor sufficient protections from genetic discrimination. Law professor Karen Rothenberg and National Center for Human Genome Research Institute staff member Elizabeth Thomson point out that a great deal of literature exists on the biological safety of prenatal genetic technologies but little on "the psychological, sociocultural, ethical, legal or political impact of their application on women and their pregnancy experience."[129] In addition to the potential benefits of genetic information, learning one's genetic status can have negative effects on one's emotional well-being, self-concept, relationship with family members and other individuals, and insurability and employability.[130] People have been denied insurance and employment benefits because of their genetic status. Insurers and some policymakers argue that continued discrimination is justified. Their fiscal logic is similar to that used in the earlier eugenics movement—that healthy people (that is, people with "good genes") should not have to support people who have or may develop genetic diseases (people with "bad genes").

Society must face today the vexing question of how the fruits of genetic research should be used. The task of developing policy in this field is similar to that of writing science fiction. The question is, What will our society look like if one policy approach is chosen versus another? Studies of the actual impact of genetic services on individuals, groups, and society at large help us to evaluate the alternative futures that genetics may bring and guide individuals and society through the choices that genetics raises.

Notes

1. Genetics Enters Our Lives

1. Their results were reported in Jeffrey P. Struewing, Dvorah Abeliovich, Tamar Peretz, Naaman Avishai, Michael M. Kaback, Francis S. Collins, and Lawrence C. Brody, "The Carrier Frequency of the BRCA1 185delAG Mutation Is Approximately 1 Percent in Ashkenazi Jewish Individuals," 11 *Nature Genet.* 198–200 (October 1995).
2. James Watson, *The Double Helix* (New York: Macmillan, 1968).
3. Susan Jenks, "Breast Cancer Gene Found," 86 *J. Nat'l Cancer Inst.* 1444–1445 (1994).
4. Struewing et al., "The Carrier Frequency," 5. Bernadine Healy, "BRCA Genes: Bookmaking, Fortunetelling, and Medical Care," 336 *N. Eng. J. Med.* 1448–1449 (1997).
5. Bernadine Healy, "BRCA Genes: Bookmaking, Fortunetelling, and Medical Care," 336 *N. Eng. J. Med.* 1448–1449 (1997).
6. L. M. McConnell, B. A. Koenig, H. T. Greely, T. A. Raffin, and the Alzheimer's Disease Working Group of the Stanford Program on Genomics, Ethics, and Society, "Genetic Testing and Alzheimer's Disease: Has the Time Come?" 4 *Nature Medicine* 757–759, 758 (1998).
7. Marcelle Morrison Bogorad, Creighton Phelps, and Neil Buckholtz, "Alzheimer Disease Research Comes of Age," 277 *J.A.M.A.* 837–840 (1997).
8. Ming-Xin Tang, Yaakov Stern, Karen Marder, Karen Bell, Barry Gurland, Rafael Lantigua, Howard Andrews, Lin Feng, Benjamin Tycko, and Richard Mayeux, "The APOE-ε 4 Allele and the Risk of Alzheimer Disease Among African

Americans, Whites, and Hispanics," 279 *J.A.M.A.* 751–755 (1998). A. M. Saunders, C. Hulette, K. A. Welsh-Bohmer, D. E. Schmechel, B. Crain, J. R. Burke, M. J. Alberts, W. J. Strittmatter, J. C. S. Breitner, C. Rosenberg, S. V. Scott, P. C. Gaskell Jr., M. A. Pericak-Vance, and A. D. Roses, "Specificity, Sensitivity, and Predictive Value of Apolipoprotein-E Genotyping for Sporadic Alzheimer's Disease," 348 *Lancet* 90 (1996), point out: "The epidemiological studies needed are incomplete, and accurate predictions of risk specific for age, sex, race or ethnic origin are not yet possible."

9.   An example of a disorder that most people consider to be serious is Tay-Sachs disease, which generally causes a painful death by age three.

10. Bryan Christie, "The Human Map," *Scotsman*, May 13, 1996, at 14.

11. R. A. C. Roos, M. Vegter-van der Vlis, J. Hermons, H. M. Elshove, A. C. Moll, J. J. P. van de Kamp, and G. W. Bruyn, "Age at Onset in Huntington's Disease: Effect of Line of Inheritance and Patient's Sex," 28 *J. Med. Genet.* 515–519 (1991).

12. See chapters 3 and 7.

13. Andrew Baum, Andrea L. Friedman, and Sandra G. Zakowski, "Stress and Genetic Testing for Disease Risk," 16 *Health Psychol.* 8–19, 9 (1997).

14. Jeffrey P. Struewing, Patricia Hartge, Sholom Wacholder, Sonya M. Baker, Martha Berlin, Mary McAdams, Michelle M. Timmerman, Lawrence C. Brody, and Margaret A. Tucker, "The Risk of Cancer Associated with Specific Mutations of BRCA1 and BRCA2 Among Ashkenazi Jews," 336 *N. Eng. J. Med.* 1401–1408 (1997); Bernadine Healy, "BRCA Genes — Bookmaking, Fortunetelling, and Medical Care," 336 *N. Eng. J. Med.* 1448–1449 (1997).

15. See, e.g., Francis S. Collins, "BRCA1: Lots of Mutations, Lots of Dilemmas," 334 *N. Eng. J. Med.* 186–188, 187 (1996).

16. *Diamond v. Chakabarty*, 447 U.S. 303 (1980).

17. *Funk Bros. Seed Co. v. Kalo Inoculant Co.*, 333 U.S. 127 (1948).

18. See, e.g., Rebecca S. Eisenberg, "Patenting the Human Genome," 39 *Emory L.J.* 721 (1990).

19. 15 U.S.C.S. § 3701 et seq. (1996); 35 U.S.C. § 200 et seq. See also Sheldon Krimsky, *Biotechnics and Society* (New York: Praeger, 1991).

20. Leon Rosenberg, "Using Patient Materials for Production Development: A Dean's Perspective," 33 *Clinical Research* 412–454 (October 1985).

21. Vincent Kiernan, "Truth Is No Longer Its Own Reward," *New Scientist*, March 1, 1997, at 11.

22. Jeannie Kever, "A New Genetic Test May Enable People Prone to Periodontal Disease to Seek Early, Aggressive Treatment," *Sarasota Herald-Tribune*, April 2, 1997, at 1E; Patricia Apodaca, "Irvine Firm's Dental Lasers Win FDA's OK," *Los Angeles Times*, May 8, 1997, at A1.

23. Gina Kolata, "A Headstone, a Coffin, and Now, the DNA Bank," *New York Times*, December 24, 1996, at C1. The chief executive officer of GeneLinks notes that funeral directors are used to selling.

24. This is because linkage studies are less necessary with increasing discoveries that allow direct testing of genes.

25. Their brochure of GENE*R*ATION LINKS lists their address as 106 N. 21st Street East, Suite 215, Superior, Wisconsin 54880–6546.

26. http://www.confidentialgenetest.com.

27. Confidential medical information about International Histocompatibility Workshop tissue donors was accessible on the Internet for a period of eight months. "The primary Internet file transfer protocol site used by the IHW is located at the Centre Interuniversitaire de Calcul in Toulouse, France (ftp://ftp.cict.fr). The user name and password necessary for access to the IHW database was printed in the IHW online newsletter, 'Express,' http://web.cict.fr/12ihwc on a regular basis in an unintended act of misfeasance. . . . [T]he number of access 'hits' and the level of intrusion as of June 1996 has been compiled and is accessible at http://www.sv.cict.fr/12ihwc/news/c7dca.html." Carlos K. Poza, "The Pandora's Box of Modern Bioethics: *Moore v. Regents of the University of California*: Whose Body Is This Anyway?" at nn. 55–56 (1997) (paper on file with author).

28. For a description of the technology, see S. Elias, J. Price, M. Dockter, S. Wachtel, A. Tharapel, J. L. Simpson, and K. W. Klinger, "First Trimester Diagnosis of Trisomy 21 in Fetal Cells from Maternal Blood," 340 *Lancet* 1033 (1992); see also Jane Chuen and Mitchell S. Golbus, "Prenatal Diagnosis Using Fetal Cells from the Maternal Circulation," 159 *West. J. Med.* 308 (1993); Richard Saltus, "Noninvasive Way Is Cited to Deter Down Syndrome in Fetuses," *Boston Globe*, November 12, 1992, at 8.

29. Down syndrome is caused by extra genetic material on chromosome 21, which results in various malformations and mental retardation. See American Academy of Pediatrics, "Health Supervision of Children with Down Syndrome," 93 *Pediatrics* 855–859 (1994).

30. Cystic fibrosis is "the most common potentially fatal genetic disease" among Caucasians. "It is caused by a disorder of exocrine glands. Individuals with cystic fibrosis have a variety of physical abnormalities, most serious among them is chronic obstructive lung disease." Congress of the United States, Office of Technology Assessment, U.S. Congress, *Healthy Children: Investing in the Future*, appendix M, 263 (Washington, D.C.: U.S. Government Printing Office, 1988).

31. Tay-Sachs disease is a fatal neurodegenerative disorder caused by a genetic mutation. It is very common among Ashkenazi Jews. See, e.g., E. C. Landel, I. H. Ellis, A. H. Fensom, P. M. Green, and M. Bobrow, "Frequency of Tay-

Sachs Disease Splice and Insertion Mutations in the UK Ashkenazi Jewish Population," 28 *J. Med. Genet.* 177–180 (1991).

32. See, e.g., Lori B. Andrews, Jane E. Fullarton, Neil A. Holtzman, and Arno G. Motulsky, eds., *Assessing Genetic Risks: Implications for Health and Social Policy*, 177–178 (Washington, D.C.: National Academy Press, 1994). For a description of one form of the technology, see Jeffrey Young, "Lab on a Chip," *Forbes*, September 23, 1996, at 210.

33. *Berkeley County Dept. of Social Services v. David Galley and Kimberly Galley*, 92-DR-08–2699 (Moncks Corner, South Carolina, April 1, 1994).

34. Lori Andrews, "Genetic Privacy: From the Laboratory to the Legislature," 1 *Human Genome Research* 1 (October 1995); Paul R. Billings, Mel A. Kohn, Margaret de Cuevas, Jonathan Beckwith, Joseph S. Alper, and Marvin R. Natowicz, "Discrimination as a Consequence of Genetic Testing," 50 *Am. J. Hum. Genet.* 476–482 (1992).

35. Erik Lineala, "Renewed Debate Surfaces Around Human Genome Project," 20, no. 4 *Alternatives* 12 (September 1994); Diane E. Lewis, "Under a Genetic Cloud," *Boston Globe*, August 14, 1994, at A1.

36. Abby Lippman, "The Genetic Construction of Prenatal Testing: Choice, Consent, or Conformity for Women?" in Karen H. Rothenberg and Elizabeth Thomson, eds., *Women and Prenatal Testing: Facing the Challenges of Genetic Technology* at 9–34, 13 (Columbus: Ohio University Press, 1994).

37. Ibid. at 14.

38. "Report and Recommendations of the Panel to Assess the NIH Investment and Research on Gene Therapy," 9, 32 (Stuart H. Orkin and Arno G. Motulsky, co-chairs, December 7, 1995). Also available at http://www.nih.gov/news/panelrep.html.

39. Dorothy Nelkin, "The Social Dynamics of Genetic Testing: The Case of Fragile-X," 10, no. 4 *Medical Anthropology Quarterly* 537–550, 544 (1996). See generally Dorothy Nelkin and M. Susan Lindee, *The DNA Mystique: The Gene as Cultural Icon* (New York: Freeman, 1995).

40. P. T. Rowley, S. Loader, C. J. Sutera, and M. Walden, "Do Pregnant Women Benefit from Hemoglobinopathy Carrier Detection?" 565 *Ann. N.Y. Acad. Sci.* 152–160 (1989); Wayne Grody, "Molecular Pathology, Informed Consent, and the Paraffin Block," 4, no. 3 *Diagnostic Molecular Pathology* 155–157 (1995).

41. John Robertson's Statement to the National Bioethics Advisory Commission, March 14, 1997, at 83. See also "Human Cloning: Should the United States Legislate Against It?" 83 *A.B.A.J.* 80–81 (May 1997).

42. John C. Bailar III and Heather L. Gornick, in "Cancer Undefeated," 336 *N. Eng. J. Med.* 1569–1574 (1997), note that "the war on cancer is far from over. . . . The effect of new treatments for cancer has been largely disappointing."

43. Jeremy Rifkin and his Foundation on Economic Trends, for example, seem to have opposed any use of genetics as necessarily harmful.

44. Serge Romensky, "Human Cloning Condemned by International Health Community," *Agence France-Presse*, May 13, 1997.

45. The White House, Office of the Press Secretary, "Remarks on Completion of the First Survey of the Entire Genome," June 26, 2000, available at http://www.whitehouse.gov/library/Press Release.

2. Competing Frameworks for Genetics Policy

1. In Great Britain, the Royal College of Obstetricians and Gynecologists organized the Interim Licensing Authority to scrutinize research and clinical services involving in vitro fertilization—such as genetic testing of embryos—and determine whether such interventions should be offered at all, and, if so, whether particular doctors and clinics should be allowed to offer these services. See Derek Morgan and Robert G. Lee, *Blackstone's Guide to the Human Fertilisation and Embryology Act 1990* 83–84 (London: Blackstone, 1991). In 1991 the Human Fertilisation and Embryology Authority, a British government agency, took over supervision and licensing of research involving human embryos.

2. Royal Commission on New Reproductive Technologies, "Royal Commission on New Reproductive Technologies—Update" (December 1993). See also Royal Commission on New Reproductive Technologies, *Proceed with Care*, vols. 1 and 2 (Canada: Minister of Government Services, 1993).

3. J. Craig Venter (presentation at International Conference on Mammalian Cloning, Washington, D.C., June 26, 1997).

4. Lori B. Andrews, Jane E. Fullarton, Neil A. Holtzman, and Arno G. Motulsky, eds., *Assessing Genetic Risks: Implications for Health and Social Policy* 270 (Washington, D.C.: National Academy Press, 1994).

5. National Action Plan on Breast Cancer and NIH-DOE Working Group on Ethical, Legal, and Social Implications of Human Genome Research, Conference on Genetic Discrimination and Health Insurance: A Case Study on Breast Cancer (July 11, 1995).

6. *Cloning Human Beings: Report and Recommendations of the National Bioethics Advisory Commission* (Rockville, Md.: National Bioethics Advisory Commission, June 1997).

7. See, e.g., Office of Technology Assessment, U.S. Congress, *Human Gene Therapy* (Washington, D.C.: U.S. Government Printing Office, 1984); Office of Technology Assessment, U.S. Congress, *Genetic Monitoring and Screening in the Workplace* (Washington, D.C.: U.S. Government Printing Office, 1990).

8. C. T. Cashey, M. M. Kaback, and A. L. Beaudet, "American Society of Human Genetics Statement on Cystic Fibrosis Screening Tissue Samples," 46 *Am. J. Hum. Genet.* 393 (1990).

9. American College of Medical Genetics Storage of Genetics Materials Committee, "Statement on Storage and Use of Genetic Materials," 57 *Am. J. Hum. Genet.* 1499–1500 (1995).

10. Biotechnology and the European Public Concerted Action Group, "Europe Ambivalent on Biotechnology," 357 *Nature* 945–947, 945 (1997).

11. Ibid. at 947. The researchers note that this is in keeping with Ulrich Beck, *Risk Society: Towards a New Modernity* (London: Sage, 1992), and Anthony Giddens, *The Consequences of Modernity* (Stanford: Stanford University Press, 1990).

12. The OTA Committee, in its cost-benefit analyses, assumed in most of the scenarios that 100 percent of women with affected fetuses would abort. Office of Technology Assessment, U.S. Congress, *Cystic Fibrosis and DNA Tests: Implications of Carrier Screening* 39–40 (OTA-BA-532, Washington, D.C.: U.S. Government Printing Office, 1992). This is in sharp contrast to studies of carrier/carrier couples, which found that only 20 percent said they would abort an affected fetus. Dorothy C. Wertz, Janet M. Rosenfield, Sally R. Janes, and Richard W. Erbe, "Attitudes Toward Abortion Among Parents of Children with Cystic Fibrosis," 81 *Am. J. Pub. Health* 992 (1991).

13. An exception is Andrews et al., *Assessing Genetic Risks.*

14. National Institutes of Health Workshop on Population Screening for the Cystic Fibrosis Gene, 323 *N. Eng. J. Med.* 70–71 (1990).

15. American Society of Human Genetics and American College of Medical Genetics, "Points to Consider: Ethical, Legal, and Psychosocial Implications of Genetic Testing in Children and Adolescents," 57 *Am. J. Hum. Genet.* 1233–1241 (1995).

16. Task Force on Genetic Information and Insurance, NIH-DOE Working Group on Ethical, Legal, and Social Implications of Human Genome Research, *Genetic Information and Health Insurance* (Bethesda, Md.: National Center for Human Genome Research, 1993).

17. The biomedical policy area is rife with examples showing that policy does not get adopted until there is a visible public case — and then the policy that gets adopted is narrowly tailored to the facts of that case. In the early 1980s, for example, a number of legislatures considered proposed statutes that would have comprehensively dealt with surrogate motherhood. See, e.g., Lori B. Andrews, *New Conceptions: A Consumer's Guide to the Newest Infertility Treatments, Including In Vitro Fertilization, Artificial Insemination, and Surrogate Motherhood* 217–218 (New York: Ballantine, 1985) (describing proposed Michigan law). It was not until after the much-publicized Baby M case, however, that lawmakers began to adopt laws in this area — and those laws were merely tailored to the problems raised by that one case. See Lori Andrews, *Between Strangers: Surrogate Mothers, Expectant Fathers, and Brave New Babies* (New York: Harper and Row, 1989).

18. Marque-Luisa Miringhoff, *The Social Costs of Genetic Welfare* 25–27 (New Brunswick, N.J.: Rutgers University Press, 1991).

19. Benjamin S. Wilfond and Kathleen Nolan, "National Policy Development for the Clinical Application of Genetic Diagnostic Technologies," 270 *J.A.M.A.* 2948–2952 (1993).

20. Robert Blank, *Regulating Reproduction* 139, 180 (New York: Columbia University Press, 1990).

21. A government commission advised Secretary of Health and Human Services Donna Shalala to appoint an advisory body to consider the range of issues raised by genetics. See "Report of the Joint NIH/DOE Committee to Evaluate the Ethical, Legal, and Social Implication Programs of the Human Genome Project" (December 12, 1996). Instead, she appointed a narrow group to advise on technical aspects of genetic testing. See Press Release, "Shalala Appoints Chair and Members of Genetic Testing Advisory Committee," June 4, 1999. The "committee was chartered . . . to help the department formulate policies on the development, validation, and regulation of genetic tests, particularly DNA-based diagnostics."

22. See chapter 6.

23. Elliot Marshall, "Varmus Proposes to Scrap the RAC," 272 *Science* 945 (1996).

24. Ibid.

25. Editorial, "Supervising Gene Therapy Openly," 350 *Lancet* 79 (1997).

26. Rick Weiss and Deborah Nelson, "Victim's Dad Faults Gene Therapy Team," *Washington Post*, February 3, 2000, at A2.

27. Kenneth M. Ludmerer, *Genetics and American Society* (Baltimore: Johns Hopkins University Press, 1972).

28. Jonathan Beckwith, "Social and Political Uses of Genetics in the United States: Past and Present," 265 *Ann. N.Y. Acad. Sci.* 46, 47 (1976) (citation omitted).

29. Ibid. at 48. See also Daniel J. Kevles, *In the Name of Eugenics: Genetics and the Uses of Human Heredity* 100 (New York: Knopf, 1985).

30. Philip Reilly, "Eugenic Sterilization in the United States," in Aubrey Milunsky and George Annas, eds., *Genetics and the Law, III*, 227 (New York: Plenum, 1985).

31. Daniel J. Kevles, "Out of Eugenics: The Historical Politics of the Human Genome," in Daniel J. Kevles and Leroy Hood, eds., *The Code of Codes: Scientific and Social Issues in the Human Genome Project* 53, 101 (Cambridge: Harvard University Press, 1992).

32. Beckwith, "Social and Political Uses of Genetics" at 47. See also Kevles, *In the Name of Eugenics* at 46.

33. See Kevles, *In the Name of Eugenics* at 9 (citation omitted).

34. Historian Daniel Kevles points out that "class and race prejudice were pervasive in eugenics." Ibid.

35. Beckwith, "Social and Political Uses of Genetics" at 48.
36. See discussion in chapter 5.
37. The *New York Times*, a respectable newspaper, asserted at the time: "Demonstrations were always mobs composed of foreign scum, beer-smelling Germans, ignorant Bohemians, uncouth Poles and wild-eyed Russians." Beckwith, "Social and Political Uses of Genetics" at 47 (citation omitted). Beckwith points out that "the eugenics movement served mainly as an ideological weapon against the poorer classes in society that were seeking a greater share in wealth and power." Ibid. at 50.
38. Ibid. at 48 (citation omitted).
39. Lori Andrews, *Medical Genetics: A Legal Frontier* 12, 13 (Chicago: American Bar Foundation, 1987).
40. Reilly, "Eugenic Sterilization" at 228.
41. Kevles, *In the Name of Eugenics* at 93.
42. Ibid. at 72–73.
43. Ibid. at 62.
44. Reilly, "Eugenic Sterilization" at 230–231.
45. Philip Reilly, *The Surgical Solution: A History of Involuntary Sterilization in the United States* 94 (Baltimore: Johns Hopkins University Press, 1991).
46. In 1923 Gustav Boeters, the leading eugenics advocate in Germany, had acknowledged the American influence: "What we racial hygienicists promote is not at all new or unheard of. In a cultured nation of the first order, in the United States of America, that which we strive toward was introduced and tested long ago." Beckwith, "Social and Political Uses of Genetics" at 49, citing Boeters, *Die Unfuchbarmachung der Geistig Minder Wertigen* (July 9–11, 1923). In 1936 Harry Laughlin, the head of the Eugenics Records Office in the United States, enthusiastically accepted an honorary doctorate from Heidelberg University for his critical role in shaping the law that was the model for the Germans. Garland E. Allen, "The Eugenics Age Revisited," 99 *Technology Review* 22–31, 26 (August/September 1996).
47. Reilly, "Eugenic Sterilization" at 228.
48. Ibid. at 227, 235.
49. Kevles, *In the Name of Eugenics* at 114.
50. Andrews, *Medical Genetics* at 12. In the early twentieth century, "a number of scientists and social scientists applied Darwinian analysis to various 'racial' groups and decided that some 'races' were more advanced than others on the evolutionary scale." Mark H. Haller, *Eugenics: Hereditarian Attitudes in American Thought*, x (New Brunswick, N.J.: Rutgers University Press, 1984).
51. Beckwith, "Social and Political Uses of Genetics" at 49.
52. Ibid. at 47.
53. Reilly, "Eugenic Sterilization" at 229.

54. Beckwith, "Social and Political Uses of Genetics" at 49.

55. Stephen Jay Gould, *The Mismeasure of Man* 166 (New York: Norton, 1981).

56. Kevles, *In the Name of Eugenics* at 49, quoting Charles Davenport, *Heredity in Relation to Eugenics* (New York: Holt, 1911).

57. Ibid. at 97 (footnotes omitted).

58. Neil A. Holtzman, personal communication, November 1, 1998.

59. Message from the Minister in New Reproductive and Genetic Technologies, *Setting Boundaries, Enhancing Health* (Government of Canada, June 1996), available at http://www.hc-sc.gc.ca/english/nrgt.

60. *Salgo v. Leland Stanford Jr. University Board of Trustees*, 154 Cal. App. 2d 560, 317 P.2d 170 (1957).

61. Lori B. Andrews, "Informed Consent Statutes and the Decisionmaking Process," 5 *J. Legal Med.* 163–217 (1984).

62. Jay Katz, *The Silent World of the Doctor/Patient Relationship* (New York: Free Press, 1984), points out that "disclosure and consent, except in the most rudimentary fashion, are obligations alien to medical thinking and practice" at 1.

63. Charles Lidz and Alan Meisel, "Informed Consent and the Structure of Medical Care," in President's Commission for the Study of the Ethical Problems in Medicine and Biomedical and Behavioral Research, *Making Health Care Decisions: The Ethical and Legal Implications of Informed Consent in the Patient-Practitioner Relationship*, 2:399–405 (Washington, D.C.: U.S. Government Printing Office, 1982).

64. In one survey, 88 percent of physicians believed that "patients want doctors to choose for them the best alternative." In contrast, 72 percent of the public said they wanted the decisions to be made jointly. President's Commission for the Study of Ethical Problems in Medicine and Biomedical and Behavioral Research, *Making Health Care Decisions: The Ethical and Legal Implications of Informed Consent in the Patient-Practitioner Relationship*, 1:46 (1982).

65. Reported in "Behavioral Medicine," *Denver Rocky Mountain News*, September 6, 1998, at 1F.

66. Ibid.

67. Andrews, *Medical Genetics* at 77. The most famous decision using the reasonable patient standard is *Canterbury v. Spence*, 464 F.2d 772 (D.C. Cir.), *cert. denied*, 409 U.S. 1064 (1972).

68. *Helling v. Carey*, 83 Wash.2d 514, 519 P.2d 981 (1974) held that in the unique situation of a physician's failure to offer a glaucoma test to a patient under forty, "the standard of care for the specialty of glaucoma was inadequate to protect the plaintiff from the incidence of glaucoma." This, however, is one of only a handful of such cases. Moreover, immediately after the case, the Washington legislature tried to avoid such cases in the future by passing a statute, Wash Rev. Code chapter 4.24.290, requiring that plaintiffs in malpractice actions prove that the physician had violated the medical standard of care.

69. See, e.g., Patricia A. Baird and Charles R. Scriver, "Genetics and Public Health," in John M. Last and Robert B. Wallace, eds., *Public Health and Preventive Medicine* 983 (13th ed., East Norwalk, Conn.: Appleton and Lange, 1991).

70. *Jacobson v. Massachusetts*, 197 U.S. 11 (1905). "The State legislature proceeded upon the theory which recognized vaccination as at least an effective if not best known way in which to meet and suppress the evils of a smallpox epidemic that imperiled an entire population." Ibid. at 28.

71. Daniel F. Roses, "From Hunter and the Great Pox to Jenner and Smallpox," 175 *Surg. Gynecol. Obstet.* 365–372 (1992).

72. *Jacobson v. Massachusetts*, 197 U.S. 11, 29 (1905).

73. Janet Gallagher, "Prenatal Invasions and Interventions: What's Wrong with Fetal Rights," 10 *Harv. Women's L.J.* 9 (1986).

74. *McFall v. Shimp*, 10 Pa. D.&C. 3d 90 (1980), as cited in James Childress, "Artificial and Transplanted Organs," 1 *BioLaw* § 13, 303, 312 (1986).

75. See *In re George*, 625 S.W.2d 151 (Mo. App. 1981), *on remand*, 630 S.W.2d 614 (Mo. App. 1982).

76. *Curran v. Bosze*, 141 Ill.2d 473, 153 Ill. Dec. 213, 566 N.E.2d 319 (1990).

77. *Head v. Colloton and Filer*, 331 N.W.2d 870 (Iowa 1983).

78. "CDC Speaks Out on Genetic Testing," *Medical Utilization Management*, May 28, 1998.

79. For a summary of some of these approaches, see Lori Andrews, "Public Choices and Private Choices: Legal Regulation of Genetic Testing," in Marc Lappé and Timothy Murphy, eds., *Justice and the Human Genome Project* 46–74 (Berkeley and Los Angeles: University of California Press, 1994).

80. There is currently variable coverage of genetic testing and counseling by state Medicaid programs. Office of Technology Assessment, *Cystic Fibrosis and DNA Tests* (forty-five state Medicaid programs cover amniocentesis and twenty-six state Medicaid programs cover DNA analysis).

81. Reilly, *The Surgical Solution* at 94.

82. Andrews, *Medical Genetics* at 18.

83. Ibid. at 238.

84. Ibid.

85. N. Holtzman, "Dietary Treatment for Inborn Errors of Metabolism," 21 *Annual Review of Medicine* 335–356 (1970).

86. See Philip Reilly, *Genetics, Law, and Social Policy* 62–86 (Cambridge: Harvard University Press, 1977).

87. See, e.g., *Planned Parenthood of Southeastern Pennsylvania v. Casey*, 505 U.S. 833 (1992). The U.S. Supreme Court in *Casey* said: "For two decades of economic and social developments people have organized intimate relationships and made choices that define their views of themselves and their places in

society, in reliance on the availability of abortion in the event that contraception should fail. The ability of women to participate equally in the economic and social life of the Nation has been facilitated by their ability to control their reproductive lives." Ibid. at 856.

88. See, e.g., 42 U.S.C.S. § 300a-5, 42 C.F.R. § 50.203–.205 (Michie 1997), and 42 C.F.R. § 441.250–.259 (Michie 1997). Family planning sterilizations performed with federally assisted funding are permissible only after obtaining non-coerced, voluntary, and informed consent from competent patients. Health care providers in federally assisted family planning projects are prohibited from performing sterilization operations on individuals from whom they did not obtain informed consent. The federal guidelines include specific provisions that must be met when acquiring informed consent prior to the procedure. The federal requirements exist to ensure that people who undergo sterilization procedures do so voluntarily and knowingly.

89. 42 U.S.C. § 263a-1(a). Virginia has a similar law. Va. Code Ann. 54.1–2971.1.

90. See, e.g., Va. Code Ann. § 20–160.

91. See, e.g., La. R.S. § 9:128.

92. Va. Code Ann. § 54.1–2971.1.

93. Americans with Disabilities Act, 42 U.S.C. § 12182(a).

94. Under the Illinois Human Rights Act, 775 ILCS 5/2–101, for example, an employer may not discriminate against an employee or applicant based on age, marital status, sex, religion, or national origin.

95. Under the City of Chicago Human Rights Ordinance, 2–160–030, employers may not discriminate against employees or applicants based on race, color, sex, age, religion, disability, national origin, ancestry, sexual orientation, marital status, parental status, military discharge status, or source of income.

96. Under the existing fundamental rights analysis with respect to abortion, the U.S. Supreme Court has taken the position that federal constitutional law does not require the public funding of abortion. However, some state legislatures have enacted laws funding abortion for poor women, and some courts have held that the state constitution requires funding of abortion to enable women to exercise their fundamental right to privacy to make reproductive decisions. See, e.g., *Moe v. Secretary of Admin.* 417 N.E.2d 387 (Mass. 1981).

97. 735 F.Supp. 1361 (N.D. Ill. 1990), *aff'd without opinion, sub nom., Scholberg v. Lifchez*, 914 F.2d 260 (7th Cir. 1990), *cert. denied*, 498 U.S. 1069 (1991).

98. Mass. Ann. Laws. ch. 112 § 12J; Mich. Comp. Laws. § 333.2685 to 2692; N.H. Rev. Stat. Ann. sec 168-B:15(II); N.D. Cent. Code § 14–02.2–01; R.I. Gen. Laws. § 11–54–1; Utah Code Ann. 76–7–310.

99. *Becker v. Schwartz*, 386 N.E.2d 807, 814 (N.Y. 1978).

100. See discussion of statutes dealing with insurance and employment discrimination in chapter 7. See also Council on Ethical and Judicial Affairs, "Use of Genetic Testing by Employers," 266 J.A.M.A. 1827 (1991).

3. *The Impact of Genetic Services on Personal Life*

1. M. Lipkin, L. Fisher, P. T. Rowley, S. Loader, and H. P. Iker, "Genetic Counseling of Asymptomatic Carriers in a Primary Care Setting," 105 *Ann. Intern. Med.* 115–123 (1986).

2. Barton Childs, Leon Gordis, Michael M. Kaback, and Haig Kazazian Jr., "Tay-Sachs Screening: Social and Psychological Impact," 28 *Am. J. Hum. Genet.* 550–558 (1976); Eila K. Watson, Edward S. Mayall, Judith Lamb, Jean Chapple, and Robert Williamson, "Psychological and Social Consequences of Community Carrier Screening Programme for Cystic Fibrosis," 340 *Lancet* 217–220, 218 (1992); Susan Zeesman, Carol L. Clow, Lola Cartier, and Charles H. Scriver, "A Private View of Heterozygotes: Eight-Year Follow-up Study on Carriers of the Tay-Sachs Gene Detected by High School Screening in Montreal," 18 *Am. J. Med. Genet.* 769–778, 772 (1984).

3. Watson et al., "Psychological and Social Consequences" at 218.

4. Zeesman et al., "A Private View of Heterozygotes" at 772.

5. Theresa M. Marteau, "Psychological Implications of Genetic Screening," 28 *Birth Defects: Original Article Series* 185–190, 185 (1992). (Although Tay-Sachs carriers viewed their current health status no differently than noncarriers, carriers' perception of future health and risk of illness was significantly more negative than that of noncarriers.)

6. Ora Gilber and Riva Borovik, "How Daughters of Women with Breast Cancer Cope with the Threat of the Illness," 24 *Behavioral Medicine* 115–121 (1998).

7. Robert T. Croyle, Ken R. Smith, Jeffrey R. Botkin, Bonnie Baty, and Jean Nash, "Psychological Responses to BRCA1 Mutation Testing: Preliminary Findings," 16 *Health Psych.* 63–72 (1997).

8. K. M. Kask, J. C. Holland, M. S. Halper, and D. G. Miller, "Psychological Distress and Surveillance Behaviors of Women with a Family History of Breast Cancer," 84 *J. Nat'l Cancer Institute* 27–30 (1992).

9. One woman felt she needn't worry about breast cancer because "it's my mother's side of the family that has breast cancer . . . and . . . I take after my father's side." G. Geller, M. Strauss, B. A. Bernhardt, N. A. Holtzman, " 'Decoding' Informed Consent: Insights from Women Regarding Breast Cancer Susceptibility Testing," 27, no. 2 *Hastings Center Report* 28–33, 29 (1997).

10. Leslie G. Bluman, Barbara K. Rimer, Donald A. Berry, Nancy Borstelmann, J. Dirk Iglehart, Katherine Regan, Joellen Schildkraut, and Eric P. Winer, "Attitudes, Knowledge, and Risk Perceptions of Women with Breast and/or Ovarian Cancer Considering Testing for BRCA1 and BRCA2," 17, no. 3 *Journal of Clinical Oncology* 1040–1046 (1999).

11. Ibid. at 1043.

12. Ibid.

13. A. C. DudokdeWit, A. Tibben, H. J. Duivenvoorden, P. G. Frets, M. W. Zoe-
    teweij, M. Losekoot, A. van Haeringen, M. F. Niermeljer, and J. Passchier,
    "Psychological Distress in Applicants for Predictive DNA Testing for Autosomal
    Dominant, Heritable Late Onset Disorders," 34 *Am. J. Med. Genet.* 382–90,
    382 (1997).
14. Ibid.
15. Ibid. at 386.
16. Richard Saltus, "Genetic Clairvoyance," *Boston Globe*, January 8, 1995, mag-
    azine at 14.
17. From the program from her Rockford College Exhibit, March 19–April 24,
    1993.
18. Jo Revill, "Why I Had a Mastectomy Before Cancer Was Diagnosed," *Evening
    Standard*, December 1, 1993, at 12. Wilson ultimately had a preventive mas-
    tectomy.
19. At any given time, about 25,000 Americans are suffering from Huntington's
    disease, while at the same time 150,000 others live knowing that they have a
    50 percent chance of having inherited the gene and thus may develop the
    disease. See Peter Gorner, "Out of the Shadow a New Genetic Test Can Fore-
    tell Agonizing Death: Would You Take It?" *Chicago Tribune*, August 4, 1988,
    at C1.
20. Lindsay A. Farrer, "Suicide and Attempted Suicide in Huntington's Disease:
    Implications for Preclinical Testing of Persons at Risk," 24 *Am. J. Med. Genet.*
    305 (1986).
21. Maurice Bloch, Shelin Adam, Sandy Wiggins, Marlene Huggins, and Michael
    Hayden, "Predictive Testing for Huntington's Disease in Canada: The Expe-
    rience of Those Receiving an Increased Risk," 42 *Am. J. Med. Genet.* 499–507,
    504 (1992).
22. Sally W. Vernon, Ellen R. Gritz, Susan K. Peterson, Christopher I. Amos,
    Catherine A. Perz, Walter F. Baile, and Patrick M. Lynch, "Correlates of Psy-
    chologic Distress in Colorectal Cancer Patients Undergoing Genetic Testing
    for Hereditary Colon Cancer," 16 *Health Psychol.* 73–86 (1997).
23. A. Tibben, H. J. Duivenvoorden, M. Vegter-van der Vlis, M. F. Niermeljer, P.
    G. Frets, J. J. van de Kamp, R. A. Roos, H. G. Rooijmans, and F. Verhage,
    "Presymptomatic DNA testing for Huntington Disease: Identifying the Need
    for Psychological Intervention" 48, no. 3 *Amer. J. Med. Genet.* 137–144 (1993).
24. Andrew Baum, Andrea L. Friedman, and Sandra G. Zakowski, "Stress and
    Genetic Testing for Disease Risk," 16 *Health Psychol.* 8–19, 8 (1997).
25. T. M. Marteau and E. Anionwu, "Evaluating Carrier Testing: Objectives and
    Outcomes," in T. M. Marteau and M. P. M. Richards, eds., *The Troubled Helix:
    Social and Psychological Implications of the New Human Genetics* 123–139
    (Cambridge, Eng.: Cambridge University Press, 1996).

26. H. Bekker, G. Dennis, M. Moddel, M. Bobrow, and T. Marteau, "The Impact of Screening for Carriers of Cystic Fibrosis," 31 *J. Med. Genet.* 364–368 (1994).

27. Robert T. Croyle, Ken R. Smith, Jeffrey R. Botkin, Bonnie Baty, and Jean Nash, "Psychological Responses to BRCA1 Mutation Testing: Preliminary Findings," 16 *Health Psychol.* 63–72, 69, 70 (1997).

28. Saltus, "Genetic Clairvoyance" at 14.

29. Studies finding that negative results do not necessarily reduce stress include C. Lerman, B. Trock, B. K. Rimer, C. Jepson, D. Brody, and A. Boyce, "Psychological Side Effects of Breast Cancer Screening," 10 *Health Psychol.* 259–267 (1991); H. T. Lynch and P. Watson, "Genetic Counselling and Hereditary Breast/Ovarian Cancer," 339 *Lancet* 1181 (letter) (1992).

30. Kimberly Quaid, J. Brandt, R. R. Faden, and S. E. Folstein, "Knowledge, Attitude, and the Decision to Be Tested for Huntington's Disease," 36 *Clinical Genetics* 431–438 (1989); Nancy S. Wexler, "Presymptomatic Testing for Huntington's Disease: Harbinger for the New Genetics," in Z. Bankowski and A. M. Capron, eds., *Genetics, Ethics, and Human Values: Human Genome Mapping, Genetic Screening and Gene Therapy, Proceedings of the XXIVth CIOMS Conference Z* (based on the meeting, held under the auspices of the Science Council of Japan, cosponsored by WHO and UNESCO, and held in Tokyo and Inuyama City, Japan, July 22–27, 1990) (Geneva: Council for International Organizations of Medical Sciences, 1991).

31. Saltus, "Genetic Clairvoyance" at 14.

32. Marlene Huggins, Maurice Bloch, Sandi Wiggins, Shelin Adam, Oksana Suchowersky, Michael Trew, Mary Lou Klimek, Cheryl R. Greenberg, Michael Eleff, Louise P. Thompson, Julie Knight, Patrick MacLeod, Kathleen Girard, Jane Theilmann, Amy Hedrick, and Michael R. Hayden, "Predictive Testing for Huntington Disease in Canada: Adverse Effects and Unexpected Results in Those Receiving a Decreased Risk," 42 *Am. J. Med. Genet.* 508–515, 508 (1992).

33. If one of their parents has the disorder, there is a 50 percent chance that they will inherit the genetic mutation and get the disorder themselves.

34. Huggins et al., "Predictive Testing for Huntington Disease in Canada" at 510.

35. Ibid.

36. Theresa Marteau and Martin Richards, eds., *The Troubled Helix: Social and Psychological Implications of the New Human Genetics* (Cambridge, Eng.: Cambridge University Press, 1996). Alice Wexler, for example, assumed that since she shared many of her mother's physical traits (such as her dark hair) she, rather than her golden-haired sister, Nancy, would inherit Huntington's disease. Alice Wexler, *Mapping Fate: A Memoir of a Family, Risk, and Genetic Research* (Berkeley: University of California Press, 1996).

37. Aad Tibben, Reiner Tinman, Erna C. Bannick, and Hugo Duivenvoorden, "Three-Year Follow-up After Presymptomatic Testing for Huntington's Disease in Tested Individuals in Partners," 16 *Health Psychol.* 20–35, 32 (1997).
38. Ibid.
39. L. B. Jakobsen, U. Malt, B. Nilsson, S. Rosenlund, and A. Heiberg, "Psychological Consequences of Presymptomatic Genetic Testing," 119 *Tidsskr Nor Laegeforen* 1913–1916 (1999).
40. Deborah J. MacDonald, "Genetic Predisposition Testing for Cancer: Effects on Families' Lives," 12 *Holistic Nursing Practice* 9 (April 1998).
41. Laurie Garrett, "A Hidden Killer in Cajun Country," *Newsday*, November 26, 1990, at 4.
42. Benedict Carey, "Chance of a Lifetime," 8 *Health* 90 (1994).
43. See chapter 5.
44. Eric Juengst, "The Perils of Genetic Genealogy," 10 *CenterViews* 1 (Winter 1996) (publication of Center for Biomedical Ethics, Case Western Reserve University School of Medicine); Arthur L. Caplan, "Handle with Care: Race, Class, and Genetics," in Timothy F. Murphy and Marc A. Lappé, eds., *Justice and the Human Genome Project* 30–45 (Berkeley: University of California Press, 1994).
45. Denise Grady, "Who Is Aaron's Heir? Father Doesn't Always Know Best," *New York Times,* January 19, 1997, at 4.
46. Elizabeth Neus, "'Genetic Research Should Not Lead to Discrimination,' Researchers Say," Gannett News Service, April 22, 1998.
47. Dean Hamer and Peter Copeland, *The Science of Desire* (New York: Simon and Schuster, 1995).
48. II. G. Brunner, M. Nelen, X. O. Brakefield, H. H. Ropers and B. A. van Oost, "Abnormal Behavior Associated with a Point Mutation in the Structural Gene for Monoamine Oxidase A," 262 *Science* 578 (October 22, 1993).
49. Nicola Barry, "Parents of Law-Abiding Children with an Abnormal Gene Deny That It Is 'Criminal,'" *Scotland on Sunday*, August 4, 1996, at 6.
50. See, e.g., M. Lipkin, L. Fisher, P. T. Rowley, S. Loader, and H. P. Iker, "Genetic Counseling of Asymptomatic Carriers in a Primary Care Setting," 105 *Ann. Intern. Med.* 115–123 (1986). There are also speculative philosophical writings about how genetic technologies might change self-concept. See, e.g., Dan W. Brock, "The Human Genome Project and Human Identity," 29 *Hous. L. Rev.* 7–22 (1992).
51. Nancy Wexler, "Genetic Jeopardy and the New Clairvoyance," 6 *Progress in Medical Genetics* 277, 300 (1985).
52. Linda Saslow, "Genetic Tests Gaining for Adult-Onset Ills," *New York Times*, May 3, 1998, at 19.
53. Office of Technology Assessment, U.S. Congress, *Healthy Children: Investing in the Future* 263 (Washington, D.C.: U.S. Government Printing Office, 1988).

54. "Cystic Fibrosis Phase I/II Clinical Trial with Gene-Based Therapeutic Commences," *Disease Weekly Plus*, June 23, 1997; Melissa A. Rosenfeld and Francis S. Collins, "Gene Therapy for Cystic Fibrosis," 109 *Chest* 241 (1996).

55. Benjamin S. Wilfond and Norman Fost, "The Introduction of Cystic Fibrosis Carrier Screening Into Clinical Practice: Policy Considerations," 70 *Milbank Quarterly* 629–659, 640 (1992).

56. Jeffrey R. Botkin and Sonia Alemagno, "Carrier Screening for Cystic Fibrosis: A Pilot Study of the Attitudes of Pregnant Women," 82 *Am. J. Pub. Health* 723–725, 724 (1992).

57. Ellen Wright Clayton, Vickie L. Hannig, Jean P. Pfotenhanes, Robert A. Parker, Preston W. Campbell III, and John A. Phillips III, "Lack of Interest by Nonpregnant Couples in Population-Based Cystic Fibrosis Carrier Screening," 58 *Am. J. Hum. Genet.* 617–627 (1996).

58. Ellen S. Tambor, Barbara A. Bernhardt, Gary A. Chase, Ruth R. Faden, Gail Geller, Karen J. Hoffman, and Neil A. Holtzman, "Offering Cystic Fibrosis Carrier Screening to an HMO Population: Factors Associated with Utilization," 6 *Am J. Hum. Genet.* 626–637 (1994).

59. J. F. Gusella, N. S. Wexler, M. Conneally, S. L. Naylor, M. A. Anderson, R. E. Tanzi, P. C. Watkins, K. Ottina, M. R. Wallace, A. Y. Sakaguchi, A. B. Young, I. Shoulson, E. Bonilla, and J. B. Martin, "A Polymorphic DNA Marker Genetically Linked to Huntington's Disease," 306 *Nature* 234–238 (1983).

60. Tibben et al., "Three-Year Follow-up" at 20.

61. Huntington's Disease Collaborative Research Group, "A Novel Gene Containing a Trinucleotide Repeat That Is Expanded and Unstable in Huntington's Disease Chromosomes," 72, no. 6 *Cell* 971–983 (1993).

62. One reason they assumed it would be beneficial was because, no matter what the result, it would relieve uncertainty. But uncertainty is sometimes beneficial. In some cases, uncertainty itself can reduce distress. Andrew Baum, Andrea L. Friedman, and Sandra G. Zakowski, "Stress and Genetic Testing for Disease Risk," 16 *Health Psychol.* 8–19, 12 (1997).

63. S. Kessler, "Psychiatric Implications of Presymptomatic Testing for Huntington's Disease," 7 *Am. J. of Orthopsychiatry* 212–219 (1987).

64. M. Bloch, M. Fahy, S. Fox, and M. R. Hayden, "Predictive Testing for Huntington's Disease: II. Demographic Characteristics, Life-Style Patterns, Attitudes, and Psychosocial Assessments of the First Fifty-one Test Candidates," 32 *Am. J. Med. Genet.* 217, 222 (1989). See also D. Craufurd, A. Dodge, L. Kerzin-Storrar, and R. Harris, "Uptake of Presymptomatic Testing of Huntington's Disease," 2 *Lancet* 603, 604 (1989).

65. Robert T. Croyle and Caryn Lerman, "Interest in Testing for Colon Cancer Susceptibility: Cognitive and Emotional Correlates," 22, no. 2 *Preventive Medicine: An Interpersonal Journal Devoted to Practice and Theory* 284–292 (1993).

66. Gail Geller, Barbara A. Bernhardt, Kathy Helzlsouer, Neil A. Holtzman, Michael Stefanek, and Patti M. Wilcox, "Informed Consent and BRCA1 Testing," 11 *Nature Genet.* 364, 364 (December 1995).

67. Ibid. at 364.

68. Andrew Baum, Andrea L. Friedman, and Sandra G. Zakowski, "Stress and Genetic Testing for Disease Risk," 16 *Health Psychol.* 8–19, 17 (1997) (references omitted). The authors note that a study of 140 women with family histories of breast cancer found that cancer worries were associated with poorer adherence to mammography screening. *Ibid.* at 16, citing C. Lerman, B. Rimer, B. Trock, A. Balshem, and P. F. Engstrom, "Factors Associated with Repeat Adherence to Breast Cancer Screening," 19 *Prev. Med.* 279–290 (1990). Another study found that the women at highest risk of breast cancer performed fewer self-examinations and those with close relatives diagnosed with breast cancer were less likely to undergo mammograms. Lerman et al., "Psychological Side Effects of Breast Cancer Screening."

69. Baum et al., "Stress and Genetic Testing" at 10.

70. Tibben et al., "Presymptomatic DNA Testing for Huntington's Disease."

71. Jo Revill, "Why I Had a Mastectomy Before Cancer Was Diagnosed," *Evening Standard*, December 1, 1993, at 12. Wilson had a preventive mastectomy. Similarly, after Cheryl Corin-Bonder's mother, grandmother, and aunt got breast cancer, she says, "I looked in the mirror every day and I couldn't stand my breasts. I felt they would kill me. I wanted to save my life, and I didn't care what anyone thought." Carol Ann Campbell, "Cheating Cancer Fears Drives Some to Surgery Before Disease Hits," *Bergen Record*, May 15, 1994, at A01. Dr. T. S. Ravikumar, codirector of the Comprehensive Breast Care Cancer Center in New Brunswick, estimates that several hundred American women have prophylactic mastectomies each year. Ibid. Yet, since not all women with a breast cancer gene mutation will get cancer, some women have had mastectomies unnecessarily.

72. Charles Siebert, "The DNA We've Been Dealt," *New York Times Magazine*, September 17, 1995, at sec. 6, p. 50 (quote from Barbara Biesecker, codirector of the National Institutes of Health's genetic counseling program).

73. Douglas Birch, "The Mattingly Mystery," *Baltimore Sun*, January 15, 1995, at 6.

74. Croyle et al., "Psychological Responses to BRCA1 Mutation Testing" at 65.

75. J. K. M. Gevers, "Genetic Testing: The Legal Position of Relatives of Test Subjects," 7 *Med. Law.* 161–166, 163 (1988).

76. See Dorothy C. Wertz and John C. Fletcher, *Ethics and Human Genetics: A Cross-Cultural Perspective* (1989); D. C. Wertz and J. C. Fletcher, "An International Survey of Attitudes of Medical Geneticists Toward Mass Screening and Access to Results," 104 *Public Health Reports* 35–44 (1989).

77. A recent study of 639 women with a family history of breast cancer who underwent prophylactic mastectomy between 1960 and 1993 found that the procedure significantly reduced their expected incidence of breast cancer. Lynn C. Hartmann, Daniel J. Schaid, John E. Woods, Thomas P. Crotty, Jeffrey L. Meyers, P. G. Arnold, Paul M. Petty, Thomas A. Sellers, Joanne L. Johnson, Shannon K. McDonnell, Marlene H. Frost, and Robert B. Jenkins, "Efficacy of Bilateral Prophylactic Mastectomy in Women with a Family History of Breast Cancer," 340 N. Eng. J. Med. 77–84 (1999). However, the study did not address whether women who had a mutation in a breast cancer gene would be benefited by prophylactic mastectomy. The problem in the latter case is this: if women with a mutation who never would have gotten breast cancer have their breasts removed, those cases will erroneously be reported as "successes" of the mastectomy procedure, rather than as personal tragedies for women who have had their breasts removed unnecessarily.

78. A Canadian group studying predictive testing in HD reported a case study of such a woman who received an increased-risk result. Bloch et al., "Predictive Testing for Huntington Disease in Canada" at 501.

79. A. C. DudokdeWit et al., "BRCA1 in the Family."

80. Ibid. at 63.

81. Ibid. at 68.

82. Ibid.

83. Pru Irvine, "I Was the First Mother in the Family to Have a Mentally Handicapped Child. Or Was I?" The Independent (London), May 7, 1997, at 10.

84. Jeffrey Obser, "Genetic Pressures/Experts Are Pushing More Testing for Disease," Newsday, February 24, 1998, at C03.

85. Ibid.

86. See Wertz and Fletcher, Ethics and Human Genetics; Wertz and Fletcher, "An International Survey."

87. Bloch et al., "Predictive Testing for Huntington's Disease: II. Demographic Characteristics" at 222.

88. Gina Kolata, "A Headstone, a Coffin, and Now, the DNA Bank," New York Times, December 24, 1996, at C1.

89. Diane Beeson, personal communication.

90. Saltus, "Genetic Clairvoyance" at 14.

91. This is discussed in Eric Juengst, "The Ethics of Prediction: Genetic Risk and the Physician-Patient Relationship," 1 Genome Science and Technology 21–36, 25 (1995), citing J. G. Compton, J. J. DiGiovanna, S. K. Santucci, K. S. Kearns, C. I. Amos, D. L. Abangan, B. P. Korge, O. W. McBride, P. M. Steinert, and S. J. Bale, "Linkage of Epidermolytic Hyperkeratosis to the Type II Keratin Gene Cluster on Chromosome 12q," 1 Nature Genet. 301–305 (1992).

92. J. H. Fanos and J. P. Johnson, "CF Carrier Status: The Importance of Not Knowing," 55, no. 3 Am. J. Hum. Genet. A292 (#1711) (1994).

93. Ibid. at 54.
94. Douglas M. Birch, "Geneticists Hunt Illness Affecting Maryland Family," *Baltimore Sun*, March 3, 1998, at 1A.
95. Birch, "The Mattingly Mystery," at 6.
96. Lori Andrews, *Medical Genetics: A Legal Frontier* 12 (Chicago: American Bar Foundation, 1987).
97. Daniel Kevles, *In the Name of Eugenics: Genetics and the Uses of Human Heredity* 54 (New York: Knopf, 1985).
98. Ibid. at 56.
99. See, e.g., "Minimal Genetic Screening for Gamete Donors," 46 *Fertility and Sterility* 83S (supp. 1, 1996).
100. Ohio Rev. Stat. Ann. § 3111.33 (requires "appropriate" screening, which "may include, but are not limited to . . . Karyotyping, . . . Tay-Sachs, sickle cell").
101. Beverly Merz, "Matchmaking Scheme Solves Tay-Sachs Problem," 258 *J.A.M.A.* 2636 (November 20, 1987).
102. For example, Amish groups exhibit distinctive genetic disorders at a much higher rate than is seen in non-Amish groups. The Amish of Lancaster County, Pennsylvania, have children with congenital recessive deafness at a rate four times that of the normal prevalence in non-Amish communities, and they have children with phenylketonuria at a rate of one in 62 versus the one in 25,000 rate that is seen in non-Amish communities.
103. In fact, 80 percent of the Amish in Lancaster County have the same eight last names. See Brian E. Albrecht, "Loving Handicapped Kids: The Amish Way," *Plain Dealer*, September 11, 1994, at 8.
104. Ibid.
105. Lori Andrews, *New Conceptions: A Consumer's Guide to the Newest Infertility Treatments, Including In Vitro Fertilization, Artificial Insemination, and Surrogate Motherhood* 79 (New York: Ballantine, 1985).
106. In one case, a woman at risk of Huntington's disease married a man with a genetic disorder. Later, when she was tested and found she didn't inherit the gene, she divorced him. Once she learned that she was "normal," she felt she deserved someone better.
107. Timothy S. Rooney, "Family Learning to Cope with Fragile X Syndrome" *Chicago Daily Herald*, October 1, 1997, at 6.
108. Nicole M. Resnick, "Absolutely No Regrets," *Pittsburgh Post-Gazette*, March 17, 1998, at G2.
109. Botkin and Alemagno, "Carrier Screening for Cystic Fibrosis" at 725.
110. Zeesman et al., "A Private View of Heterozygotes" at 773.
111. B. Childs et al., "Tay-Sachs Screening" at 552.
112. George Stamatoyannopoulous, "Problems with Screening and Counseling in the Hemoglobinopathies," in A. G. Motulsky and F. J. G. Ebling, eds., *Birth*

Defects: Proceedings of the Fourth International Conference 268–276 (Amsterdam: Excerpta Medica, 1974).

113. Ibid. at 274.

114. Ibid.

115. D. K. Wellisch, E. R. Gritz, W. Schain, H. J. Wang, and J. Siau, "Psychological Functioning of Daughters of Breast Cancer Patients. I: Daughters and Comparison Subjects," 32 *Psychosomatics* 324–336 (1991), as cited in Caryn Lerman and Marc Schwartz, "Adherence and Psychological Adjustment Among Women at High Risk of Breast Cancer," 28 *Breast Can. Res. and Treat.* 145–155, 149 (1993). Although this study was conducted on women with an increased risk of breast cancer, the data reported are useful in determining the effects of genetic testing. Genetic testing often can determine if an individual will develop or is at an increased risk of developing disease; however, testing often falls short of identifying the severity of disease manifestation or the time of disease onset. This uncertainty is very similar to the uncertainty experienced by individuals at high risk of disease because of family history. Many of the same thoughts and fears that affect the daily functioning of daughters of breast cancer patients may be present in individuals who receive genetic test results that leave uncertainties.

116. A Canadian study assessed changes in relationships for individuals participating in predictive testing for Huntington's disease by administering questionnaires before test results were received and then again at 7–10 days, 6, 12, 18, and 24 months after the receipt of the results. The study included 217 individuals: 53 received an increased-risk result, 96 received a decreased-risk result, 33 received uninformative results, and 35 ultimately decided not to participate in testing. There were no differences between groups as far as gender, employment status, marital/relationship status, or number of children. However, the mean age of the decreased-risk group was significantly older than the mean age of the other groups. T. Copley, C. M. Benjamin, S. Wiggins, W. McKellin, S. Adam, M. Bloch, J. L. Theilmann, S. Cox, M. R. Hayden, and Canadian Collaborative Study for P.T., "Significant Changes in Social Relations After Predictive Testing for Huntington Disease," 55, no. 3 *Am. J. Hum. Genet.* A291 (#1707) (1994). A U.S. study of individuals participating in predictive testing for Huntington's disease found similar results. Nineteen couples participated in testing. Five of the couples received increased-risk results. All five reported higher levels of marital stress twelve months after receipt of test results than did individuals with decreased-risk results. Kimberly A. Quaid and Melissa K. Wesson, "The Effects for Predictive Testing for Huntington Disease on Intimate Relationships," 55, no. 3 *Am. J. Hum. Genet.* A294 (#1728) (1994).

117. Jakobsen et al., "Psychological Consequences of Presymptomatic Genetic Testing."

118. Michael Huggins, Maurice Bloch, S. Wiggins, S. Adam, O. Suchowersky, M. Trew, M. Klimek, C. R. Greenberg, M. Eleff, L. P. Thompson, J. Knight, Patrick MacLeod, Kathleen Girard, Jane Thielmann, Amy Hedrick, and Michael R. Hayden, "Predictive Testing for Huntington Disease in Canada" at 511.

119. Ibid. at 513–514.

120. Saltus, "Genetic Clairvoyance" at 14.

121. M. B. Hans and A. H. Koeppen, "Huntington's Chorea: Its Impact on the Spouse," 168, no. 4 *J. Nervous and Mental Disease* 209–214 (1980).

122. "Once a Dark Secret," 308 *Brit. Med. J.* 542 (1994); Brendan P. Minogue, Robert Taraszweski, Sherman Elias, and George J. Annas, "The Whole Truth and Nothing But the Truth?" 18, no. 5 *Hastings Center Report* 34 (1988).

123. Ann-Marie Codori, Philip R. Slavney, Candace Young, Diana L. Miglioretti, and Jason Brandt, "Predictors of Psychological Adjustment to Genetic Testing for Huntington's Disease," 16 *Health Psychol.* 36–50, 46 (1997).

124. See Gorner, "Out of the Shadow."

4. The Changing Face of Parenthood in the Genetics Era

1.   http://www.genetest.com.

2.   Stephen G. Post, Jeffrey R. Botkin, and Peter Whitehouse, "Selective Abortion for Familial Alzheimer Disease," 79, no. 5 *Obstet. and Gynecol.* 794–798 (1992).

3.   C. Tamura, N. Kakee, and Y. Tsunematsu, "Questionnaire Regarding Genetic Testing of Cancer Answered by Cancer Children's Families" (abstract, National Society of Genetic Counselors 1999 Education Conference).

4.   Post, Botkin, and Whitehouse, "Selective Abortion" at 794.

5.   Ibid. at 795.

6.   *Lifchez v. Hartigan*, 735 F.Supp. 1361, 1377 (N.D. Ill. 1990), *aff'd without opinion, sub nom., Scholberg v. Lifchez*, 914 F.2d 260 (7th Cir. 1990), *cert. denied* 498 U.S. 1069 (1991).

7.   See Lori Andrews, "Torts and the Double Helix: Malpractice Liability for Failure to Warn of Genetic Risks," 23 *Hous. L. Rev.* 149 (1992). See, e.g., *Becker v. Schwartz*, 386 N.E.2d 807 (N.Y. 1978) (finding liability where a reasonable health care provider should have known of the risk of Down syndrome because of the mother's advanced age).

8.   Ibid. See, e.g., *Curlender v. Bio-Science Lab.*, 165 Cal. Rptr. 477 (Ct. App. 1980) (suit against a laboratory for negligently conducting Tay-Sachs genetic testing).

9.   Genetic counselors point out that "no matter how glad they are that they terminated the pregnancy or how much they love the health-impaired child they

have, they still will be saddened by the loss of the child or the things the child cannot do." M. Malinowski, "Coming Into Being: Law, Ethics, and the Practice of Prenatal Screening," 45 *Hastings L.J.* 1435–1526, 1467 (1994).

10. Abby Lippman, "The Genetic Construction of Prenatal Testing: Choice, Consent, or Conformity for Women?" in Karen H. Rothenberg and Elizabeth J. Thomson, eds., *Women and Prenatal Testing: Facing the Challenges of Genetic Technology* 9–34 (Columbus: Ohio State University Press, 1994).

11. Ibid. at 26.

12. Office of Technology Assessment, U.S. Congress, *Human Gene Therapy— Background Paper,* appendix A at 64 (Washington, D.C.: U.S. Government Printing Office, 1984).

13. See also M. S. Golbus, W. D. Loughman, C. J. Epstein, G. Habasch, J. D. Stephens, B. D. Hall, "Prenatal Diagnosis in 3000 Amniocenteses," 300 *N. Eng. J. Med.* 157–163 (1979).

14. "Chorionic Villus Sampling and Subsequent Abortion; Adapted from Obstetrics and Gynecology, May 1994; Tips from Other Journals," 50 *American Family Physician* 1368 (1994).

15. "ACOG Opinion on Chorionic Villus Sampling: American College of Obstetricians and Gynecologists," 53 *American Family Physician* 1895 (1996).

16. Mark E. Deutchman and Ellen Louise Sakornbut, "Diagnostic Ultrasound in Labor and Delivery," 51 *American Family Physician* 145 (1995).

17. P. G. Pryde, A. Drugan, M. P. Johnson, N. B. Isada, and M. I. Evans, "Prenatal Diagnosis: Choices Women Make About Pursuing Testing and Acting on Abnormal Results," 36 *Clinical Obstetrics and Gynecology* 469–509 (1993).

18. Lippman, "The Genetic Construction of Prenatal Testing."

19. Barbara Katz Rothman, *The Tentative Pregnancy* (New York: Viking Penguin, 1986). Katz Rothman interviewed sixty women who underwent amniocentesis, as well as sixty who refused it. Women who had amniocentesis began wearing maternity clothes later than those who refused testing and were less likely to announce their pregnancy until later in their term. They were also less likely to remember when they first felt the fetus move. At 98.

20. B. Dixon, T. L. Richards, R. N. Reinsch, B. S. Edrich, M. R. Matson, and O. W. Jones, "Mid-trimester Amniocentesis: Subjective Maternal Responses," 26 *J. Repro. Med.* 10–16 (1981).

21. Ruth Hubbard, *The Politics of Women's Biology* 195 (New Brunswick, N.J.: Rutgers University Press, 1990).

22. Neural tube defects are serious birth defects of the brain or spinal cord resulting from failure of closure of the covering of the brain or spinal cord.

23. Rose Green [pseud.], "Letter to a Genetic Counselor," 1 *J. Genet. Counsel.* 55–70, 58 (1992).

24. Elena A. Gates, "The Impact of Prenatal Testing on Quality of Life in Women," 8 *Fetal Diagnosis and Therapy* 236–243, 240 (supp. 1) (1993).

25. Dorothy C. Wertz, "How Parents of Affected Children View Selective Abortion," in Helen Bequaert Holmes, ed., *Reproductive Technology I: An Anthology* 161–189, 176 (New York: Garland, 1992).
26. Ibid. at 176–177.
27. Adrienne Asch, "Reproductive Technology and Disability," in Sherrill Cohen and Nadine Taub, eds., *Reproductive Laws for the 1990's* 69–124, 83 (Clifton, N.J.: Humana, 1989).
28. Lippman, "The Genetic Construction of Prenatal Testing."
29. Eleanor Singer, "Public Attitudes Towards Genetic Testing," 10 *Population Research and Policy Review* 235–255 (1991).
30. Audrey Heimler, "Group Counseling for Couples Who Have Terminated a Pregnancy Following Prenatal Diagnosis," 26 *Birth Defects; Abstracts of Selected Articles* 161–167, 167 (1990).
31. Green, "Letter to a Genetic Counselor" at 61–62.
32. Rayna Rapp, "Refusing Prenatal Diagnosis: The Uneven Meaning of Bioscience in a Multicultural World," 23 *Science, Technology, and Human Values* 45–70 (1998).
33. Heimler, "Group Counseling" at 165.
34. Green, "Letter to a Genetic Counselor" at 67.
35. Heimler, "Group Counseling" at 166.
36. Rita Black, "Reproductive Genetic Testing and Pregnancy Loss: The Experience of Women," in Karen H. Rothenberg and Elizabeth J. Thomson, eds., *Women and Prenatal Testing: Facing the Challenges of Genetic Technology* 271–294 (Columbus: Ohio State University Press, 1994).
37. Angus Clarke, "Is Non-Directive Genetic Counselling Possible?" 338 *Lancet* 998–1001, 999 (October 19, 1991).
38. Donna Gore Olsen, "Parental Adjustment to a Child with Genetic Disease: One Parent's Reflections," 23 *J. Obstet. Gynecol. Neonatal Nurs.* 516–518, 516 (1994).
39. Clarke, "Is Non-Directive Genetic Counselling Possible?" at 1000: "I contend that an offer of prenatal diagnosis implies a recommendation to accept that offer, which in turn entails a tacit recommendation to terminate a pregnancy if it is found to show any abnormality. I believe that this sequence is present irrespective of the counsellor's wishes, thoughts, or feelings, because it arises from the social context rather than from the personalities involved—although naturally the counsellor may reinforce these factors."
40. Sherman Elias and George Annas, "Generic Consent for Genetic Screening," 330 *N. Eng. J. Med.* 1611–1613, 1613 (1994).
41. Clarke, "Is Non-Directive Genetic Counselling Possible?" at 1000 (October 19, 1991).

42. See Jean Seligmann and Donna Foote, "Whose Baby Is It Anyway?" *Newsweek*, October 28, 1991, at 73; Charlotte Allen, "Boys Only: Pennsylvania's Anti-Abortion Law," 206 *New Republic* 16 (March 9, 1992).

43. Lori Andrews, "Body Science," 83 A.B.A.J. 44–49, 49 (April 1997).

44. As Dawn Johnsen notes, "In order to avoid being 'caught' by the authorities, she might not seek any prenatal care, thereby endangering both her own and her future child's health." Dawn Johnsen, "The Creation of Fetal Rights: Conflicts with Women's Constitutional Rights to Liberty, Privacy, and Equal Protection," 95 *Yale L.J.* 577, 612 n. 56 (1986).

45. Because the stress carried in coerced medical interventions during pregnancy may harm the fetus, Elizabeth Taylor suggests that a fetus may have a psychological and physical interest in his or her mother's autonomy. Note, "Constitutional Limitations in State Intervention in Prenatal Care," 67 *Va. L. Rev.* 1051, 1066 (1981).

46. M. E. Mennie, M. E. Compton, A. Gilfillan, W. A. Liston, I. Pullen, D. A. Whyte, and D. J. H. Brock, "Prenatal Screening for Cystic Fibrosis: Psychological Effects on Carriers and Their Partners," 30, no. 7 *J. Med. Genet.* 543–548, 547 (1993); M. M. Burgess, "Ethical Issues in Prenatal Testing," 27, no. 2 *Clin. Biochem.* 87–91, 89 (1994).

47. S. Adam, S. Wiggins, P. Whyte, M. Bloch, M. H. K. Shokeir, H. Solton, W. Meschino, A. Summers, O. Suchowersky, J. P. Welch, J. Theilmann, and M. R. Hayden, "Five Year Study of Prenatal Testing for Huntington's Disease: Demand, Attitudes, and Psychological Assessment," 30, 7 *J. Med. Genet.* 549–556, 552 (1993).

48. Rothman, *The Tentative Pregnancy* at 129.

49. *Curlender v. Bio-Science Laboratories*, 165 Cal. Rptr. 477 (Ct. App. 1980); *Procanik v. Cillo*, 478 A.2d 755 (N.J. 1984); and *Harbeson v. Parke-Davis*, 655 P.2d 483 (Wash. 1983).

50. 106 Cal. App. 3d 811, 829, 165 Cal. Rptr. 477 (2d Dist. 1980).

51. Nancy Wexler, "Clairvoyance and Caution: Repercussions from the Human Genome Project," in Daniel J. Kevles and Leroy Hood, eds., *The Code of Codes: Scientific and Social Issues in the Human Genome Project* 211–243, 233 (Cambridge: Harvard University Press, 1992).

52. D. C. Wertz, "Patients' Ethical Views on Genetic Privacy, Disclosure to Relatives, and Testing Children: A Pilot Study," 55, no. 3 *Am. J. Hum. Genet.* A295 (#1734) (1994).

53. Tim Beardsley, "Vital Data," *Scientific American* 100–105, 104 (March 1996).

54. Gina Kolata, "Tests to Assess Risks for Cancer Raising Questions," *New York Times*, March 27, 1995, at A1.

55. The expectations and motivations of a sixteen-year-old girl who wanted to know whether she had a genetic mutation predisposing to breast cancer are discussed

in Terence Monmaney, "Genetic Testing: Kids' Latest Rite of Passage," 9 *Health* 46 (January 1995).

56. American Society of Human Genetics and American College of Medical Genetics, "Points to Consider: Ethical, Legal, and Psychosocial Implications of Genetic Testing in Children and Adolescents," 57 *Am. J. Hum. Genet.* 1233–1241 (1995).

57. Dorothy C. Wertz, Joanna H. Fanos, and Philip R. Reilly, "Genetic Testing for Children and Adolescents: Who Decides?," 272 *J.A.M.A.* 875, 878 (1994).

58. Ibid. Similarly, the ASHG/ACMG statement notes: "Expectations of others for education, social relationships and/or employment may be significantly altered when a child is found to carry a gene associated with a late-onset disease or susceptibility. Such individuals may not be encouraged to reach their full potential, or they may have difficulty obtaining education or employment if their risk for early death or disability is revealed." ASGH/ACMG, "Points to Consider" at 1236.

59. Ibid. (citations omitted).

60. Neil A. Holtzman, "What Drives Neonatal Screening Programs?" 325 *N. Eng. J. Med.* 802–809 (September 12, 1991), referring to K. B. Hammond, S. H. Abman, R. J. Sokol, and F. J. Accurso, "Efficacy of Statewide Newborn Screening for Cystic Fibrosis by Assay of Trypsinogen Concentration," 325 *N. Eng. J. Med.* 769–774 (1991).

61. Ibid.

62. Ibid., citing P. Farrell, personal communication.

63. Statement of Claude LaBerge at June 1992 meeting of the Institute of Medicine's Committee on Assessing Genetic Risks.

64. For example, some women whose fetuses are at risk for Huntington's disease reject abortion because they believe that since the age of onset is often in the fifties, a cure will be found before the child develops the disease.

65. See chapter 3.

66. A. C. Codori, G. M. Peterson, P. A. Boyd, J. Brandt, F. M. Giadiello, "Genetic Testing for Cancer in Children: Short-Term Psychological Effect," 150 *Arch. Pediatr. Adolesc. Med.* 1131–1138, 1132 (1996).

67. Monmaney, "Genetic Testing."

68. Diane E. Hoffman and Eric Wulfsberg, "Testing Children for Genetic Predispositions: Is It in Their Best Interest?" 23 *J. Law, Medicine, and Ethics* 331–44, 333 (1995).

69. Ibid. at 333–334.

70. Norman Fost, "Genetic Diagnosis and Treatment: Ethical Considerations," 147 *Am. J. Dis. Child.* 1190–1195, 1193 (1993).

71. See chapter 3.

72. ASHG/ACMG, "Points to Consider" at 1233.

73. Hoffman and Wulfsberg, "Testing Children for Genetic Predispositions" at 333, citing Holtzman, "What Drives Neonatal Screening Programs?" at 802–804.

74. National Advisory Council for Human Genome Research, "Statement on Use of DNA Testing for Presymptomatic Identification of Cancer Risks," 271 J.A.M.A. 785 (1994). See also "Statement of the American Society of Human Genetics on Genetic Testing for Breast and Ovarian Cancer Predisposition," 55 Am. J. Hum. Genet. i–iv (1994).

75. ASHG/ACMG, "Points to Consider" at 1236.

76. Gerry Evers-Kiebooms, "Decision Making in Huntington's Disease and Cystic Fibrosis," 23, no. 2 Birth Defects: Original Article Series 115–149, 143, 144 (1987).

77. Allyn McConkie-Rosell, Gail A. Spiridigliozzi, Kathleen Rounds, Deborah V. Dawson, Jennifer A. Sullivan, Deby Burgess, and Ave M. Lachiewicz, "Parental Attitudes Regarding Carrier Testing in Children at Risk for Fragile X Syndrome," 82 Am. J. Med. Genet. 206–211 (1999).

78. Ibid. at 207.

79. Ibid. at 208. Among the same parents, 81 percent (52/64) of them were either somewhat concerned or very concerned that their child might experience discrimination if found to be a carrier. Sixty-nine percent (44/64) were not concerned that their child would be discriminated against because of his or her carrier status; however, most parents (88 percent—56/64) were either very concerned or somewhat concerned with helping their child adjust to being a carrier.

80. Hoffman and Wulfsberg, "Testing Children for Genetic Predispositions" at 332.

81. Wertz, Fanos, and Reilly, "Genetic Testing for Children and Adolescents" at 878.

82. Lori B. Andrews, Jane E. Fullarton, Neil A. Holtzman, and Arno G. Motulsky, eds., Assessing Genetic Risks: Implications for Health and Social Policy, 276 (Washington, D.C.: National Academy Press, 1994). For other analyses of the appropriate uses of genetic testing in children, see Wertz, Fanos, and Reilly, "Genetic Testing for Children and Adolescents"; ASGH/ACMG, "Points to Consider"; Hoffman and Wulfsberg, "Testing Children for Genetic Predispositions."

83. L. Went, "Ethical Issues Policy Statement on Huntington's Disease Molecular Genetics Predictive Test," 27 J. Med. Genet. 34–38 (1990).

84. Scientific Advisory Board of the National Kidney Foundation, "Gene Testing in Autosomal Dominant Adult Polycystic Kidney Disease: Results of a National Kidney Foundation Workshop," 13 Am. J. Kidney Dis. 85–87 (1989).

85. Wertz, Fanos, and Reilly, "Genetic Testing for Children and Adolescents" at 875.

86. Ibid., citing *The Genetic Testing of Children: Report of a Working Party of the Clinical Genetics Society*, Angus Clarke, Chair (Edgbaston and Birmingham, U.K.: Clinical Genetics Society, Birmingham Maternity Hospital, 1993).

87. M. Z. Pelias, "Duty to Disclose in Medical Genetics: A Legal Perspective," 39 *Am. J. Hum. Genet.* 347–354 (1991).

88. P. S. Harper and A. Clarke, "Should We Test Children for 'Adult' Genetic Diseases?," 335 *Lancet* 1205–1206 (1990).

89. Report of a Working Party of the Clinical Genetics Society (UK), "The Genetic Testing of Children," 31 *J. Med. Genet.* 785–797 (1994). The working party sent questionnaires to 35 molecular genetics labs and 23 cytogenetics labs to assess whether labs were performing genetic tests on children and, if so, what kind of tests were administered. Of the 35 molecular genetics labs, 16 responded to the questionnaire. Ten of the 23 cytogenetics labs responded to the questionnaire.

Responses of the molecular genetics labs:
- Six labs indicated that they regularly performed carrier testing on children for recessive disorders.
- Several labs indicated that they performed predictive testing in children for adenomatous polyposis, myotonic dystrophy, and adult polycystic kidney disease.
- A few labs indicated that they performed testing for X-linked disorders in children.
- Two labs indicated that they had a specific laboratory policy not to perform predictive testing or carrier testing on children.

Responses by the cytogenetics labs were not reported in a detailed manner by the authors, but they did report, generally, that a "substantial" number of genetic tests were performed on children by these labs (795).

90. Ibid.

91. Ibid. at 796.

92. Hoffman and Wulfsberg, "Testing Children for Genetic Predispositions" at 331, give the example of a manufacturer's letter to dermatologists about testing children for melanoma predisposition.

93. See Ellen Wright Clayton, "Removing the Shadow of Law from the Debate About Genetic Testing of Children," 57 *Am. J. Med. Genet.* 630 (1995).

94. Wertz, Fanos, and Reilly, "Genetic Testing for Children and Adolescents" at 880 (1994).

95. ASHG/ACMG, "Points to Consider" at 1239.

96. McConkie-Rosell et al., "Parental Attitudes Regarding Carrier Testing in Children at Risk for Fragile X Syndrome."

97. Ibid. at 208.

98. Ibid. at 209.

99. *In re Green*, 448 Pa. 338, 292 A.2d 387, 392 (1972); see, e.g., Rowine H. Brown and Richard B. Truitt, "The Right of Minors to Medical Treatment," 28 *DePaul L. Rev.* 289, 292, 299 (1979); Comment, "Who Speaks for the Child and What Are His Rights?" 4 *Law and Hum. Behav.* 217, 223 (1980).

100. Comment, "The Right to Refuse Medical Treatment: Under What Circumstances Does It Exist?" 18 *Duq. L. Rev.* 607, 613, 614–615 (1980). See, e.g., W. Wadlington, "Medical Decision Making for and by Children: Tensions Between Parent, State, and Child," 1994 *U. Ill. L. Rev.* 311, 319 (1994).

101. Hoffman and Wulfsberg, "Testing Children for Genetic Predispositions" at 334.

102. Clayton, "Removing the Shadow of the Law from Debate" at 631.

5. The Impact of Genetic Services on Women, People of Color, and Individuals with Disabilities

1.   See Anna C. Mastroianni, Ruth Faden, and Daniel Federman, eds., *Women and Health Research: Ethical and Legal Issues of Including Women in Clinical Studies* (Washington, D.C.: National Academy Press, 1994).

2.   See Robert Mendelsohn, *MalePractice: How Doctors Manipulate Women* (Chicago: Contemporary Books, 1981); Karen M. Hicks, ed., *Misdiagnosis: Woman as a Disease* (Allentown, Pa: People's Medical Society, 1994).

3.   Ruth Macklin, "Women's Health: An Ethical Perspective," 21 *J. Law, Medicine, and Ethics* 23–29, 24 (Spring 1993).

4.   Rebecca Dresser, "Wanted: Single, White Male for Medical Research," *Hastings Center Report* 24–29, 24 (January–February 1992).

5.   Ibid.; Karen H. Rothenberg, "Gender Matters: Implications for Research and Women's Health Care," 32 *Hous. L. Rev.* 1201, 1207 (1996).

6.   Dresser, "Wanted."

7.   Ibid. at 24, citing Bruce Nussbaum, "Under-the-Counter Drug Testing," *New York Times*, February 19, 1991; Paul Cotton, "Is There Still Too Much Extrapolation from Data on Middle-Aged White Men?" 263 *J.A.M.A.* 1049–1050 (1990); Kathleen Nolan, "AIDS and Pediatric Research," 14 *Evaluation Review* 464–481 (1990); Greg Sachs and Christine Cassel, "Biomedical Research Involving Older Human Subjects," 18 *Law, Medicine, and Health Care* 234–243 (1990).

8.   Macklin, "Women's Health" at 26.

9.   Ibid. at 26, citing Vicki Ratner, Debra Slade, and Kristene E. Whitmore, "Interstitial Cystitis: A Bladder Disease Finds Legitimacy," 1 *Journal of Women's Health* 63–68 (1992).

10. Dresser, "Wanted" at 25.

11. R. Alta Charo, "Protecting Us to Death: Women, Pregnancy, and Clinical Research Trials," 38 *St. Louis U.L.J.* 135–167, 150 (1992), citing 58 Fed. Reg. 39,406–39,407 (1993).

12. Mastroianni, Faden, and Federman, *Women and Health Research*, 1:135.

13. Susan Jenks, "Role for Pregnant Women in Clinical Trials Debated," 86 *National Cancer Institute* 1829–1832 (1994).

14. Gary Lee, "Patients Unaware of Role in Medical Experiments," *Chicago Sun-Times*, June 25, 1995, at 24.

15. R. Levine, *Ethics and Regulations of Clinical Research* 53 (Baltimore: Urban and Schwarzenberg, 1981).

16. Gena Corea, *The Mother Machine* 101–103 and 135 n. 2 (New York: Harper and Row, 1985).

17. Ibid. at 167.

18. Mary Anne Bobinski, "Women and HIV: A Gender-Based Analysis of a Disease and Its Legal Regulation," 3 *Tex. J. Women and Law* 7, 16 (1994).

19. Marian C. Limacher, M.D., "The Clinical Picture—Women and Heart Disease" (paper presented at Women's Health for the 21st Century conference, San Diego, California, September 20, 1997).

20. Natalie Davis Spingarn, "Women's Ill Treatment by Doctors: Authors Indict Male-Controlled Medicine for Blindness, Bias, and Boorishness," *Washington Post*, January 31, 1995, at Z13, 1995 WL 2076153.

21. Rothenberg, "Gender Matters" at 1210.

22. Steven Miles and Allison August, "Courts, Gender, and the Right to Die," 18 *Law, Medicine, and Health Care* 85 (1990).

23. *Application of President and Directors of Georgetown College*, 118 App. D.C. 80, 331 F.2d 1000, *cert. denied*, 377 U.S. 978 (1964).

24. *In re Osborne*, 294 A.2d 372 (D.C. 1972).

25. Rafter notes that Richard Dugdale's 1877 study, *The Jukes*, described poor women as dwelling in "'hot beds where human maggots are spawned.'" Nicole H. Rafter, "Claims-Making and Socio-Cultural Context in the First U.S. Eugenics Campaign," 39 *Social Problems* 17–34 (February 1993) at 21, citing Richard L. Dugdale, *The Jukes: A Study in Crime, Pauperism, Disease, and Heredity* 54 (New York: G. P. Putnam's Sons, 1877).

26. Ibid., citing Josephine Shaw Lowell, Ripley Ropes, and E. W. Foster, "Report of the Committee on a Reformatory for Women," in *New York State Board of Charities, 12th Annual Report*, New York Senate Document 13, at 189 (1879).

27. "The very possibility that poor women might use their bodies unconventionally threatened the biological understanding of gender as fixed and immutable." Rafter, "Claims-Making and Socio-Cultural Context in the First U.S. Eugenics Campaign" at 25. The eugenics movement to institutionalize dependent, allegedly feebleminded fertile women can be seen in part as "the punitive reaction to changes in women's social position." Ibid. at 29. Similarly, nonfeebleminded women who got pregnant were punished by being forced to go to term under stricter abortion laws, such as the one adopted in Connecticut

in 1860, which prohibited post-quickening abortion. *Roe v. Wade*, 410 U.S. 113, 139 (1973).

28. Rafter, "Claims-Making and Socio-Cultural Context in the First U.S. Eugenics Campaign."

29. *Muller v. Oregon*, 208 U.S. 412, 421 (1908). There is evidence, however, that the laws were designed not to protect women but to protect men's jobs. Women in low-paying, traditionally female jobs like nursing were exempted from the laws. Lori Andrews, "A Delicate Condition," 13, no. 9 *Student Lawyer* 30–36, 21 (May 1985).

30. 274 U.S. 200 (1927).

31. 410 U.S. 113, 154 (1973).

32. *Buck v. Bell*, 274 U.S. 200, 207 (1927).

33. Ibid.

34. Paul A. Lombardo, "Three Generations, No Imbeciles: New Light on *Buck v. Bell*," 60 *N.Y.U. L. Rev.* 30, 52 (1985).

35. Ibid. at 53.

36. Ibid. at 62.

37. Ibid. at 55.

38. Ibid. at 42 n.8.

39. Despite the pervasive acceptance of eugenic ideals, particularly among the well-educated, another lawyer might have done a better job. In six lower courts that heard challenges to laws authorizing sterilization of the feebleminded or insane, three declared them unconstitutional on equal protection grounds, since they applied only to institutionalized persons, and three declared them unconstitutional on due process grounds, since they did not provide adequate procedural safeguards. Philip Reilly, "Eugenic Sterilization in the United States," in Aubrey Milunsky and George Annas, eds., *Genetics and the Law III* 227, 232 (New York: Plenum, 1985).

40. A. Lippman, "Prenatal Genetic Testing and Screening: Constructing Needs and Reinforcing Inequalities," 17 *Am. J. Law and Med.* 15–50 (1991); Mary Mahowald, Dana Levinson, and Christine Cassell, "The New Genetics and Women," 74 *Milbank Quarterly* 239 (1996).

41. Only in a small minority of cases will the fetus's status exclusively provide information about the father—when the disorder is a dominant one and the gene is transmitted by the father. In other instances of dominant disorders, it is the mother who passes on the gene. In the case of recessive disorders, both parents pass on the gene. In the case of X-linked disorders, the mother passes on the genetic mutation, and 50 percent of her sons inherit it.

42. Theresa M. Marteau, Ruth Dundas, and David Axworthy, "Long-Term Cognitive and Emotional Impact of Genetic Testing for Carriers of Cystic Fibrosis: The Effects of Test Results and Gender," 16 *Health Psychol.* 51–62, 52 (1997).

43. See, e.g., P. N. Goodfellow, "Steady Steps Lead to the Gene," 341 *Nature* 102–103 (1990); D. Brock, "Population Screening for Cystic Fibrosis," 47 *Am. J. Hum. Genet.* 164–165 (1990).

44. "Statement of the American Society of Human Genetics on Cystic Fibrosis Carrier Screening," 51 *Am. J. Hum. Genet.* 1443–1444, 1444 (1992); "Statement from the National Institutes of Health Workshop on Population Screening for the Cystic Fibrosis Gene," 323 *N. Eng. J. Med.* 70–71, 71 (1990).

45. Peter T. Rowley, Starlene Loader, Jeffery Levenkron, and Charles E. Phelps, "Cystic Fibrosis Carrier Screening: Knowledge and Attitudes of Prenatal Care Providers," 9 *Am. J. Prev. Med.* 261–263 (1993); M. A. Fox, C. D. Schetter, Z. Tatsugawa, R. M. Cantor, C. Fang, J. M. Novak, H. N. Bass, F. G. Crandell, and W. W. Grody, "Consent to Cystic Fibrosis Carrier Screening in an Ethnically Diverse Population" (abstract no. 50, 1993 American Society of Human Genetics annual meeting).

46. See chapter 6.

47. Irwin M. Ellerin, Mia I. Frieder, and Gary R. Hillerich, "Handling a Failure to Diagnose Breast Cancer Case," *Trial* 31–37, 31 (May 1996), citing a Physician Insurers Association of America study.

48. See, e.g., *Pratt v. Williams*, 658 So.2d 4 (La. 1995); *Solomon v. Hall*, 767 S.W.2d 158 (Tenn. App. 1988).

49. Eight states specifically require doctors to inform women of the alternative methods of treatment (see, e.g., Cal. Health & Safety Code § 109275(a); 24 Me. Rev. Stat. § 2905-A; Md. Code Ann. § 20–113; Mass. Ann. ch. 111 § 70E; Mich. Stat. Ann. § 14.15 (MCL 17013); N.Y. Cls. Pub. Health § 2405; 35 Pa. Stat. § 5641) or require doctors to distribute brochures with that information (Ky. Rev. Stat. 311.935).

Three other statutes require that the State Department of Public Health prepare a document explaining the alternatives in lay terms for distribution to doctors and hospitals for potential distribution to patients. 20 ILCS 2310/2310–345; N.J. Stat. § 26:2–168; and Tex. Health & Safety Code § 86.001 et seq.

In some states failure to comply with the statute constitutes unprofessional conduct under the state medical licensing and disciplinary law (see, e.g., Cal. Health & Safety Code § 109275[b]). In other states, the patient may bring a civil action against the physician (see, e.g., Mass. Ann. Laws ch. 111, § 70E).

50. Dr. Frank Desposito (presentation at Seton Hall Law School, November 8, 1996).

51. Caryn Lerman and Robert Croyle, "Psychological Issues in Genetic Testing for Breast Cancer Susceptibility," 154 *Arch. Intern. Med.* 609 (March 28, 1994).

52. "It is possible that mammography may put women who carry the BRCA1 gene at increased risk of developing the disease" (testimony of Fran Visco, president, National Breast Cancer Coalition, Senate Cancer Coalition, September 29,

1995). Carriers of the A-T mutation are four to five times more likely to develop various cancers than noncarriers (testimony of Brad Margus, president, A-T Children's Project, Senate Cancer Coalition, September 29, 1995). "An estimated 2–3% of the U.S. population are carriers of the altered gene and have an increased risk for cancer. This suggests that 4–5% of all cancer (and 18% of breast cancer) in the American population may be associated with the A-T gene" (testimony of Francis Collins, Senate Cancer Coalition, September 29, 1995).

53.   Nancy Press and Carol Browner, "Collective Fictions: Similarities in the Reasons for Accepting MSAFP Screening Among Women of Diverse Ethnic and Social Class Backgrounds," 8 *Fetal Diagn. and Ther.* 97–106 (1993).

54.   P. T. Rowley, S. Loader, C. J. Sutera, and M. Walden, "Do Pregnant Women Benefit from Hemoglobinopathy Carrier Testing?" 565 *Ann. N.Y. Acad. Sci.* 152–160 (1989).

55.   J. Stein, N. Gilensky, J. Groden, and K. Huelsman, "Effects of Gender on Interest in Genetic Testing for Colorectal Cancer Among Individuals with Colorectal Cancer" (abstract, National Society of Genetic Counselors 1999 Education Conference).

56.   Marteau, Dundas, and Axworthy, "Long-Term Cognitive and Emotional Impact of Genetic Testing for Carriers of Cystic Fibrosis."

57.   Anxiety was most often associated with fears of children inheriting the disease, ignorance of the meaning of carrier status, and shock at "having been singled out." Individuals experiencing the most anxiety were women who were identified as carriers and were pregnant. B. Childs, "Tay-Sachs Screening: Social and Psychological Impact," 28 *Am. J. Hum. Genet.* 550–558, 551 (1976).

58.   Mahowald, Levinson, and Cassell, "The New Genetics and Women."

59.   Berit Sjogren, "Future Use and Development of Prenatal Diagnosis, Consumers' Attitudes," 12 *Prenatal Diagn.* 1–8, 2 (1992).

60.   No women indicated that the couple should not be the ultimate decision maker, and 18 percent were uncertain. Forty percent of male partners believed the couple should not be the decision maker in the use of prenatal diagnosis, and 40 percent were uncertain who the decision maker should be. The question proposed medical specialists or public authorities as alternative decision makers to the couple. Ibid. at 4.

61.   Marteau, Dundas, and Axworthy, "Long-Term Cognitive and Emotional Impact of Genetic Testing for Carriers of Cystic Fibrosis" at 52, citing M. Pillisuk and C. Acredolo, "Fear of Technological Hazards: One Concern or Many?" 3 *Social Behav.* 17–24 (1988); P. Slovic, "Perceptions of Risk: Reflections on the Psychometric Paradigm," in D. Goldring and S. Krimsky, eds., *Social Theories of Risk*, 117–152 (New York: Praeger, 1992).

62.   Ian Savage, "Demographic Influences on Risk Perceptions," 13 *Risk Analysis* 413–420 (1993).

63. Marc Pilisuk and Curt Acredolo, "Fear of Technological Hazards: One Concern or Many?" 3 *Social Behav.* 17–24, 23 (1988).

64. James Flynn, Paul Slovic, and C. K. Mertz, "Gender, Race, and Perceptions of Environmental Health Risks," 14 *Risk Analysis* 1101–1108, 1107 (1994).

65. Ibid.

66. Theresa M. Marteau and Harriet Drake, "Attributions for Disability: The Influence of Genetic Screening," 40 *Soc. Sci. Med.* 1127–1132 (1995).

67. B. D. Colen, "Proceedings of the Workshop on Inherited Breast Cancer in Jewish Women: Ethical, Legal, and Social Implications," 10 *CenterViews* 7–10, 9 (Spring 1996).

68. Mahowald, Levinson, and Cassel, "The New Genetics and Women."

69. *Muller v. Oregon*, 28 S. Ct. 324, 327 (1908).

70. See, e.g., *Jefferson v. Griffin Spalding City Hosp.*, 247 Ga. 86, 274 S.E.2d 457 (1987).

71. See, e.g., Margery Shaw, "Conditional Prospective Rights of the Fetus," 5 *J. Legal Med.* 63 (1984).

72. This is evident, for example, in the prosecution of women who drink alcohol or use drugs during pregnancy. See, e.g., Janet Dinsmore, *Pregnant Drug Users: The Debate Over Prosecution* (Alexandria, Va.: National Center for Prosecution of Child Abuse, American Prosecutors Research Institute, 1992).

73. Arlene Zarembka and Katherine M. Franke, "Women in the AIDS Epidemic: A Portrait of Unmet Needs," 9 *St. Louis U. Pub. L. Rev.* 519, 526 (1990). I have similarly noted a trend toward "policing pregnancy." See Andrews, "A Delicate Condition."

74. Carol Beth Barnett, "The Forgotten and the Neglected," 23 *Golden Gate U.L. Rev.* 863, 886–887 (1993) (citation omitted).

75. See, e.g., Jason Geitzen, "She's in the Army Now and Her Higher Injury Rates Concern Pentagon," *Omaha World Herald*, April 28, 1996, at 1A, in which military officials make the argument that women's greater susceptibility to stress fractures and inability "to urinate while standing in the corner of a truck bed" make them unfit for certain higher-prestige military jobs.

76. See Lisa Ikemoto, "The In/Fertile, the Too Fertile, and the Dysfertile," 47 *Hastings L.J.* 1007, 1045–1052 (1996) (discussing and refuting that argument).

77. Beverly Horsburgh, "Schrodinger's Cat, Eugenics, and the Compulsory Sterilization of Welfare Mothers: Deconstructing the Old/New Rhetoric and Constructing the Reproductive Rights of Natality for Low-Income Women of Color," 17 *Cardozo L. Rev.* 531, 535 (1996).

78. John Robert Hand, "Buying Fertility: The Constitutionality of Welfare Bonuses for Welfare Moms Who Submit to Norplant Insertion," 46 *Vand. L. Rev.* 715, 718 (April 1993). Kathleen Megan, "Proposal Offers Money for Contraceptive Use," *Hartford Courant*, February 17, 1994, at A1; Peter Mitchell, "Lawmaker

Puts Money on Birth Control Idea," *Orlando Sentinel*, February 16, 1994, at D5. See also 3 *Reproductive Freedom News*, March 11, 1994, at 5. The use of Norplant is being urged despite the fact that several thousand women have filed suit against the manufacturer with products liability complaints. Melynda G. Broomfield, "Controlling the Reproductive Rights of Impoverished Women: Is This the Way to 'Reform' Welfare?" 16 *Boston College Third World L.J.* 217, 234 n. 151 (1996).

79. In Washington, women on welfare would receive $10,000 for having a tubal litigation after one child was born. Hand, "Buying Fertility" at 718.

80. "Paying Teenagers Not to Have Babies?" *Christian Science Monitor*, January 14, 1993, People section at 14. Most of the proponents of Norplant use make the fiscal argument, reminiscent of the early advocacy by the American eugenics movement, that the rest of society should not have to support paupers whose birth could have been avoided. Some proponents use eugenic arguments as well. When David Duke, a former Ku Klux Klan Grand Wizard, proposed a Norplant bonus in Louisiana, he suggested that social problems could be averted by preventing the birth of "undesirables." Broomfield, "Controlling the Reproductive Rights of Impoverished Women" at 233.

81. *In re Baby Boy Doe*, 260 Ill. App. 3d 392, 632 N.E.2d 326 (Ill. 1994); *In re A.C.* 573 A.2d 1235 (1990).

82. Karen Lebacqz, "Feminism and Bioethics: An Overview," 17, no. 2 *Second Opinion* 11–25, 15 (October 1991).

83. Office of Technology Assessment, U.S. Congress, *Infertility: Medical and Social Choices* 9 (Washington, D.C.: U.S. Government Printing Office, 1988).

84. Va. Code Ann. § 54.1–2971 (Michie 1996).

85. 42 U.S.C. § 263a–1(a).

86. Veronika Kolder, Janet Gallagher, and Michael Parsons, "Court-Ordered Obstetrical Interventions," 316 N. Eng. J. Med. 1192–1196, 1196 (1987).

87. Lori B. Andrews, "Doctors Do Not Always Know Best," *Los Angeles Times*, January 2, 1884, at 5.

88. *In re A.C.*, 573 A.2d 1235 (D.C. Cir. 1987).

89. *In re Baby Boy Doe*, 260 Ill. App. 3d 392, 399, 632 N.E.2d 326, 330 (1994).

90. Betsy A. Lehman, "Woman Awarded $1.53 Million Because of Forced Caesarean," *Houston Chronicle*, June 27, 1993, at C6.

91. See, e.g., *In re Osborne*, 294 A.2d 372 (D.C. 1972). See also *In re Baby Boy Doe*, 632 N.E.2d 326, 330 (1994), holding that "the right to refuse treatment does not depend upon whether the treatment is perceived as risky or beneficial to the individual," citing *In re Estate of Longeway*, 549 N.E.2d 292, at 292 (1989).

92. Rowley et al., "Do Pregnant Women Benefit From Hemoglobinopathy Carrier Detection?"

93. There are financial consequences as well. The women who are being tested for sickle-cell anemia without their knowledge are paying for a test they may not want, the results of which they may not use.

94. Such a concern was raised by a pregnant woman who participated in a free research protocol at Kaiser-Permanente in California in which she was not informed in advance of the possibility that it might identify her as having two mutant cystic fibrosis genes.

95. Harriet A. Washington, "Human Guinea Pigs," *Emerge* 24–35, 28 (October 1994).

96. Ibid.

97. Arthur L. Caplan, "Handle with Care: Race, Class, and Genetics," 30–45, 32–33, in Timothy F. Murphy and Marc A. Lappé, eds., *Justice and the Human Genome Project* (Berkeley: University of California Press, 1994).

98. Barbara L. Bernier, "Class, Race, and Poverty: Medical Technologies and Sociopolitical Choices," 11 *Harvard BlackLetter Journal* 115–143, 118 (1994) (citations omitted).

99. Ibid. at 119, citing Todd L. Savitt, *Medicine and Slavery: The Disease and Health Care of Blacks in Antebellum Virginia* 293 (Urbana: University of Illinois Press, 1978).

100. Ibid.

101. Washington, "Human Guinea Pigs" at 24.

102. Bernier, "Class, Race, and Poverty" at 122 (citations omitted).

103. Cathy Cummins, "Pregnancy Suit Declared a Class Action," *Tampa Tribune*, March 28, 1996; Wayne Washington, "Hospital, USF Settle 'Dignity' Suit," *St. Petersburg Times*, March 11, 2000.

104. David J. Rothman, "Were Tuskegee and Willowbrook 'Studies in Nature'?" 12 *Hastings Center Report* 5 (April 1982); Tuskegee Syphilis Study Ad Hoc Panel to the Department of Health, Education, and Welfare, *Final Report* (1973); J. Jones, *Bad Blood: The Tuskegee Syphilis Experiment* (New York: Free Press, 1981); and T. G. Benedek, "The 'Tuskegee Study' of Syphilis: Analysis of Moral Versus Methodologic Aspects," 31 *J. Chronic Dis.* 35 (1978).

105. President William Clinton, "Remarks in Apology to African-Americans on the Tuskegee Experiment," 33 *Weekly Comp. Pres. Doc.*, May 16, 1997.

106. See, e.g., "Professor Edward M. Miller: The Newest Member of the 'Academy of Academic Affronts to Black People,'" 18 *J. Blacks Higher Educa.* (Winter 1996), available at 1996 WL 15734530 (describes a professor whose work is purportedly designed to show that "there is 'a positive correlation between head size and intelligence,'" and that "black brains are on average 100 cubic centimeters smaller than white brains"). See also John B. Winski, "There Are Skeletons in the Closet: The Repatriation of Native American Human Remains and Burial Objects," 34 *Ariz. L. Rev.* 187, 191 (1992), noting that in 1868 the

United States surgeon general instituted a crania study of Native American remains and concluded that the crania of Native Americans proved that all nonwhites were intellectually and morally inferior to whites. But see Bell Reuhlman, "Einstein's Brain, George Washington's Teeth, Truth Is Stranger Than Fiction," *Virginian-Pilot and Ledger-Star*, January 21, 1996, at J3. Reuhlman points out that Einstein's brain was recorded at 1,230 grams, "well within the normal male range of 1,200 to 1,600 . . . so the long-held theory that brain size and intelligence were somehow correlated immediately went out the window."

107. Steven Rose, "The Rise of Neurogenetic Determinism," 373 *Nature* 380–382, 381 (1995), citing V. H. Mark and F. R. Ervin, *Violence and the Brain* (New York: Harper and Row, 1970).

108. Richardson, "Legal and Ethical Issues."

109. Elizabeth Mort, Joel S. Weissman, and Arnold Epstein, "Physician Discretion and Racial Variation in the Use of Surgical Procedures," 154 *Arch. Intern. Med.* 761–767 (1994).

110. Lucette Lagnado, "When Racial Sensitivities Clash with Research," *Wall Street Journal*, June 25, 1997, at B1.

111. Ibid.

112. Ibid.

113. Reilly, "Eugenic Sterilization in the United States" at 230.

114. Jonathan Beckwith, "Social and Political Uses of Genetics in the United States: Past and Present," 265 *Ann. N.Y. Acad. Sci.* 46, 49 (1976).

115. See Lori B. Andrews, *Medical Genetics: A Legal Frontier* 18 (Chicago: American Bar Foundation, 1987).

116. Philip Reilly, *Genetics, Law, and Social Policy* 67 (Cambridge: Harvard University Press, 1977).

117. Ibid. at 74, citing J. D. Bowman, "Sickle Cell Screening: Medico-Legal, Ethical, Psychological, and Social Problems: A Sickle Cell Crisis" (paper presented at Meharry Medical College, 1972).

118. Herbert Nickens, "The Genome Project and Health Services for Minority Populations," in Thomas H. Murray, Mark A. Rothstein, and Robert F. Murray, eds., *The Human Genome Project and the Future of Health Care* 58–74, 65 (Bloomington: Indiana University Press, 1996).

119. Ibid.

120. Ibid. at 66.

121. Harriet A. Washington, "Henrietta Lacks—An Unsung Hero," *Emerge* 24–35, 29 (October 1994).

122. Nickens, "The Genome Project and Health Services" at 67.

123. This patent was later withdrawn after protest.

124. World Council of Indigenous People, "Resolution on the HGDP," Native Net Archive Page, http://broc09.uthsca.edu/natnet/archive/n1/hgdp.html.

125. Richard J. Herrnstein and Charles Murray, *The Bell Curve: Intelligence and Class Structure* in *American Life* (paperback ed., New York: Free Press, 1996). *The Bell Curve's* assertion that some individuals do not have the cognitive skill sufficient to make it in today's society is similar to the assertion by Francis Galton (who coined the term *eugenics*) that certain people in his day did not have the refined capacities to be a proper Englishman. Ibid. at 25–115.

126. For an extensive discussion of the legal and social issues raised by purported propensities to crime, see Lori Andrews, "Predicting and Punishing Antisocial Acts: How Courts Might Use Behavioral Genetics," in Ronald A. Carson and Mark A. Rothstein, eds., *Behavioral Genetics: The Clash of Culture and Biology* 116–155 (Baltimore: Johns Hopkins University Press, 1999).

127. Washington, "Human Guinea Pigs" at 33.

128. Dorothy Nelkin, "Behavioral Genetics and Dismantling the Welfare State," in Carson and Rothstein, *Behavioral Genetics* 156–171, 169. Nelkin and Lindee point out why such explanations are readily accepted by the public and by policymakers: "They can relieve personal guilt by implying compulsion, an inborn inability to resist specific behavior," and they can relieve societal guilt and give society an excuse to cut out social services by deflecting attention away from social and economic influences on behavior." Dorothy Nelkin and M. Susan Lindee, *The DNA Mystique: The Gene as Cultural Icon* (New York: Freeman, 1995).

129. Alan J. Tobin, "Amazing Grace: Sources of Phenotype Variations in Genetic Boosterism," at 8 (paper prepared for Symposium on Biology and Culture, Galveston, Texas, November 1996). Tobin points out that there is a perception that genes preordain success. "We may reasonably wonder why so many 40–60 year old male academics have been arguing so intensely about the excellence of their own genes." Ibid.

130. L. Nsiah-Jefferson, "Reproductive Genetic Services for Low-Income Women and Women of Color," in K. H. Rothenberg and E. J. Thomson, eds., *Women and Prenatal Testing: Facing the Challenges of Genetic Technology* 234–259 (Columbus: Ohio State University Press, 1994).

131. B. A. Barnhardt, G. A. Chase, R. R. Faden, G. Geller, K. J. Hofman, E. S. Tambor, and N. A. Holtzman, "Educating Patients About Cystic Fibrosis Carrier Screening in a Primary Care Setting, 5 *Arch. Fam. Med.* 336–340 (June 1996).

132. Rowley et al., "Do Pregnant Women Benefit from Hemoglobinopathy Carrier Testing?"

133. Rayna Rapp, "Refusing Prenatal Diagnosis: The Uneven Meaning of Bioscience in a Multicultural World," 23 *Science, Technology, and Human Values* 45–70 (1998).

134. Tony Pugh, "Blacks Remain Uneasy About Health Care System," *Fort Worth Star-Telegram*, August 10, 1998, at 1.

135. Gail Geller, Barbara A. Bernhardt, Kathy Helzlsouer, Neil A. Holtzman, Michael Stefanek, and Patti M. Wilcox, "Informed Consent and BRCA1 Testing," 11 *Nature Genet.* 364 (1995).

136. Dana Hawkins, "A Bloody Mess at One Federal Lab," *U.S. News and World Report*, June 23, 1997, at 26.

137. Ibid.

138. Ibid.

139. *Norman-Bloodsaw v. Lawrence Berkeley Laboratory*, 135 F.3d 1260 (9th Cir. 1998). See discussion in chapter 7.

140. Gail Vines, "Race in Black and White," 147 *New Scientist* 34–37 (July 8, 1995). But see Robert S. Boyd, "Biological Differences of Race Only Skin Deep," *Greensboro News and Record*, January 7, 1997, at D3 (quoting Yale geneticist Kenneth Kidd stating, "The DNA data support the concept that you can't draw boundaries around races").

141. See, e.g., N. O. Bianchi, G. Baillet, C. M. Bravi, R. F. Carnese, F. Rothhammer, V. L. Martinez-Marignac, and S. D. Pena, "Origin of Amerindian Y-Chromosomes as Inferred by the Analysis of Six Polymorphic Markers," 102 *Am. J. Phy. Anthropol.* 79–89 (1997); A. Torroni, Y. S. Chen, O. Semino, A. S. Santachiara-Beneceretti, C. R. Scott, M. T. Lott, M. Winter, and D. C. Wallace, "mtDNA and Y Chromosome Polymorphisms in Four Native American Populations from Southern Mexico," 54 *Am. J. Hum. Genet.* 303–318 (1994).

142. See the questions to that effect raised in Caplan, "Handle with Care" at 34–36, and in Eric Juengst, "The Perils of Genetic Genealogy," 10 *CenterViews* 1 (Winter 1996).

143. Caplan, "Handle with Care" at 32–33.

144. This problem is exacerbated by the fact that few people have contact with individuals with a disability and consequently may overestimate their cost to society and underestimate their contribution to society.

145. Adrienne Asch, for example, points out that blind parents are sometimes denied custody and visitation rights based on their disability. Adrienne Asch, "Reproductive Technology and Disability," in Sherrill Cohen and Nadine Taub, eds., *Reproductive Laws in the 1990's* 69–123, 79 (Clifton, N.J.: Humana Press, 1989).

146. Ibid. at 81.

147. Angus Clarke, "Is Non-Directive Genetic Counselling Possible?" 338 *Lancet* 998–1001, 1000 (October 19, 1991).

148. Paul Ramsey, "Screening: An Ethicist's View," 147, 159, in B. Hilton, D. Callahan, M. Harris, P. Condliffe, and B. Berkley, eds., *Ethical Issues in Human Genetics: Genetic Counseling and the Use of Genetic Knowledge* 163. Fogarty International Proceedings no. 13 (New York: Plenum Press, 1973).

149. Marsha Saxton, "Disability Rights and Selective Abortion," in Rickie Solinger, ed., *Abortion Wars: A Half-Century of Struggle, 1950 to 2000* (Berkeley: University of California Press, 1998).

150. Dorothy C. Wertz and John C. Fletcher, "A Critique of Some Feminist Challenges to Prenatal Diagnosis," 2 *J. Women's Health* 173–188, 181 (1993).

151. T. C. Brown, D. C. Wertz, R. C. Fox, and B. Lerner, "Ethical Perspectives of Genetic Counselors: Does Area of Specialty Matter?" (abstract, National Society of Genetic Counselors 1999 Education Conference).

152. Saxton, "Disability Rights and Selective Abortion."

153. Heather Draper and Ruth Chadwick, "Beware! Preimplantation Genetic Diagnosis May Solve Some Old Problems But It Also Raises New Ones," 25 *J. Med. Ethics* 114–120 (1999).

154. Ibid.

155. Laura Hershey, "Choosing Disability," *Ms.*, July/August 1994, at 26–32, 29.

156. Saxton, "Disability Rights and Selective Abortion."

157. Office of Technology Assessment, U.S. Congress, *Mapping Our Genes* 84 (Washington, D.C.: U.S. Government Printing Office, 1988).

158. Eric Juengst, "Prenatal Diagnosis and the Ethics of Uncertainty," in J. Monagle and D. Thomasma, eds., *Health Care Ethics: Critical Issues for the 21st Century* 15–29, 19 (Rockville, Md.: Aspen Publishing, 1997), citing National Institute of Child Health and Human Development, *Antenatal Diagnosis: Report of a Consensus Development Conference* 1–192, NIH Publication 79–1973 (Bethesda, Md.: NIH, 1979).

159. Diane Beeson, "Social and Ethical Issues in the Prenatal Diagnosis of Fetal Disorders," in Benedict M. Ashley and Kevin D. O'Rourke, eds., *Health Care Ethics* 76–86, 81 (St. Louis: Catholic Health Association, 1989).

160. Abby Lippman, "The Genetic Construction of Prenatal Testing, Choice, Consent, or Conformity for Women?" in Rothenberg and Thomson, *Women and Prenatal Testing* 9–34, 15.

161. Adrienne Asch and Gail Geller, "Feminism, Bioethics, and Genetics," in Susan M. Wolf, ed., *Feminism and Bioethics: Beyond Reproduction* 318–350, 330 (London: Oxford University Press, 1996).

162. In one study, "60% of urban white women over age 40 in Georgia, but only 0.5% of African women over 40 in rural areas used prenatal diagnosis." Wertz and Fletcher, "Critique of Some Feminist Challenges" at 178, citing D. C. Sokol, J. R. Byrd, A. T. L. Chen, M. F. Goldberg, G. P. Oakley, "Prenatal Chromosome Diagnosis: Racial and Geographical Variation for Older Women in Georgia," 244 *J.A.M.A.* 1355–1357 (1980).

163. See, e.g., *Gleitman v. Cosgrove*, 227 A.2d 689 (N.J. 1967); *Becker v. Schwartz*, 386 N.E.2d 807 (N.Y. 1978).

164. 44 N.J. 22, 227 A.2d 689 (1967).

165. Jonathan Swift, "A Modest Proposal" in *Gulliver's Travels and Other Writings*, 488–496 (New York: Modern Library, 1958).

166. Asch, "Reproductive Technology and Disability" at 94.

167. Ruth Faden, "Reproductive Genetic Testing, Prevention, and the Ethics of Mothering," in Rothenberg and Thomson, *Women and Prenatal Testing* 88–97, 92.

168. Ibid.

169. See Jean Seligmann and Donna Foote, "Whose Baby Is It Anyway?" *Newsweek*, October 28, 1991, at 73; Charlotte Allen, "Boys Only: Pennsylvania's Anti-Abortion Law," 206 *New Republic*, March 9, 1992, at 16.

170. Faden, "Reproductive Genetic Testing, Prevention, and the Ethics of Mothering" at 92.

171. Asch, "Reproductive Technology and Disability" at 87.

172. Hugh Gregory Gallagher, "Can We Afford Disabled People?" (Fourteenth Annual James C. Hemphill Lecture, September 7, 1995, Rehabilitation Institute of Chicago) at 21.

173. Hershey, "Choosing Disability," at 28.

174. Ibid. at 30.

175. "Disabled People Mixed Over Genetic Future," http://news.bbc.co.uk/hi/english/health/newsid_459000/459330.stm.

176. Marsha Saxton, "Disability Rights and Selective Abortion."

177. Lisa Blumberg, "Eugenics vs. Reproductive Choice," *Disability Rag and ReSource*, January/February 1994, at 5.

178. Clarke, "Is Non-Directive Counselling Possible?" at 1000.

179. About one percent of U.S. abortions occur after prenatal diagnosis. Wertz and Fletcher, "Critique of Some Feminist Challenges," citing A. Yankauer, "What Infant Mortality Tells Us," 80 *Am. J. Pub. Health* 653 (1990).

180. This could be a benefit of the Americans with Disabilities Act, which provides legal protections to assure that individuals with disabilities are better integrated into society.

181. Dorothy C. Wertz, Janet M. Rosenfield, Sally R. Janes, and Richard W. Erbe, "Attitudes Toward Abortion Among Parents of Children with Cystic Fibrosis," 81 *Am J. Pub. Health* 992 (1991).

6. Problems in the Delivery of Genetic Services

1.   Neil A. Holtzman and Michael S. Watson, eds., *Promoting Safe and Effective Genetic Testing in the United States: Final Report of the Task Force on Genetic Testing* 65 (ELSI, September 1997). http://www.nhgri.nih.gov/ELSI/TFGT Final/. Subsequent citations to this source will appear as *Final Report*.

2. Ibid.

3. E. Palmer, "Clinical Genetics in a Family Practice Residency" (abstract prepared for National Society of Genetic Counselors 1999 Education Conference).

4. P. T. Rowley, S. Loader, J. Levenkron, C. E. Phelps, "Cystic Fibrosis Carrier Screening: Knowledge and Attitudes of Prenatal Care Providers," 9 *Am. J. Prev. Med.* 261–263 (1993).

5. Neil A. Holtzman, "Primary Care Physicians as Providers of Frontline Genetic Services," 8 *Fetal Diagnosis and Therapy* 213–219 (suppl. 1, 1993).

6. George Stamatoyannopoulous, "Problems with Screening and Counseling in the Hemoglobinopathies," in A. G. Motulsky and F. J. G. Ebling, eds., *Birth Defects: Proceedings of the Fourth International Conference* 268–276, 273 (Amsterdam: Excerpta Medica, 1974).

7. *Final Report* at 63.

8. Karen J. Hofman, Ellen S. Tambor, Gary A. Chase, Gail Geller, Ruth R. Faden, and Neil A. Holtzman, "Physicians' Knowledge of Genetics and Genetic Tests," 68 *Acad. Med.* 625 (1993).

9. M. A. Kershner, E. A. Hammond, and A. E. Donnenfeld, "Knowledge of Genetics Among Residents in Obstetrics and Gynecology," 51(s) *Am. J. Hum. Genet.* A16 (1992).

10. Francis M. Giardello, Jill D. Brensinger, Gloria M. Petersen, Michael C. Luce, Linda M. Hylind, Judith A. Bacon, Susan V. Booker, Rodger D. Parker, and Stanley R. Hamilton, "The Use and Interpretation of Commercial APC Gene Testing for Familial Adenomatous Polyposis," 336 *N. Eng. J. Med.* 823–827 (1997).

11. Mark Skolnick, of Myriad Genetics (presentation at "Genetic Testing for Breast Cancer Susceptibility: The Science, the Ethics, the Future," symposium held in San Francisco, November 22, 1996).

12. Quoted in Barbara Bedway, "Are You Prepared for the New Ethical Dilemmas?" 73, no. 4 *Medical Economics* 81 (February 26, 1996).

13. http://www.nsgc.org/career_info.html, providing reprints of an October 1999 survey by the National Society of Genetic Counselors, Inc. See also Lori B. Andrews, Jane E. Fullarton, Neil A. Holtzman, and Arno G. Motulsky, eds., *Assessing Genetic Risks: Implications for Health and Social Policy*, 203, 208 (Washington, D.C.: National Academy Press, 1994).

14. Holtzman, "Primary Care Physicians." More pediatricians (78 percent) and OB/GYNs (87 percent) than primary-care physicians, internists, and psychiatrists said they offered some form of genetic counseling.

15. Andrews et al., *Assessing Genetic Risks* at 127.

16. Ibid. at 117.

17. Ibid. at 133.

18. Ibid. at 117.
19. Ibid. at 127.
20. Ibid.
21. Ibid. at 130.
22. *Final Report* at 52.
23. Memo from Tony Holtzman to Task Force on Genetic Testing, November 8, 1995 (memo discussing various counselors' assessment of quality assurance problems in genetic testing, on file with author).
24. Judy Peres, "Gene Tests Shed Light on Illness, But Also Hold Dark Side," *Chicago Tribune*, September 26, 1999, at 1, 16. Margaret M. McGovern, Marta O. Benach, Sylvan Wallenstein, Robert J. Desnick, and Richard Keenlyside, "Quality Assurance in Molecular Genetic Testing Laboratories," 281 *J.A.M.A.* 835–840 (1999).
25. Ibid.
26. C. Holtzman, W. E. Slazyk, J. F. Cordero, and W. H. Hannon, "PKU Newborn Screening: A Descriptive Epidemiology of Missed Cases of Phenylketonuria and Congenital Hypothyroidia," 78 *Pediatrics* 553–558 (1986). There are more than 5 percent false negatives in newborn screening. Andrews et al., *Assessing Genetic Risks* at 133 (citations omitted).
27. J. Vockley, J. A. Inserra, W. R. Breg, and T. L. Yang-Fang, " 'Pseudomosaicism' for 4p—in Amniotic Fluid Cell Culture Proven to Be True Mosaicism After Birth," 39 *Am. J. Med. Genet.* 81–83 (1991). See also *Martinez v. Long Island Jewish Hillside Center*, 512 N.E.2d 535 (1987).
28. Andrews et al., *Assessing Genetic Risks* at 123.
29. Giardello et al., "The Use and Interpretation of Commercial APC Gene Testing." See also Susan Gilbert, "Doctors Often Misread Results of Genetic Tests, Study Finds," *New York Times*, March 26, 1997, at C4.
30. *Curlender v. Bio-Science Laboratories*, 106 Cal. App. 3d 811, 165 Cal. Rptr. 477 (2d Dist. 1980).
31. Andrews et al., *Assessing Genetic Risks* at 118.
32. N.Y. CLS Pub. Health § 57 et seq.
33. Md. Code Ann. § 17–211.
34. Idaho Code § 5–334; Minn. Stat. Ann. § 145.424(1)–(2); Mo. Ann. Stat. § 188.130–1; N.C. Gen. Stat. § 14–45.1(e); 42 Pa. Cons. Stat. Ann. § 8305; S.D. Codified Laws Ann. § 21–55–22; Utah Code Ann. § 78–11–24.
35. Andrews et al., *Assessing Genetic Risks* at 128.
36. *Final Report* at 29–30.
37. Andrews et al., *Assessing Genetic Risks* at 129.
38. Neil A. Holtzman and Stephen Hilgartner, "Appendix 3: State of the Art of Genetic Testing in the United States: Survey of Biotechnology Laboratories and Interviews of Selected Organizations," in *Final Report* 99–124, 124 (Table 7) (September 1997).

39. Clinical Laboratories Improvement Act of 1967 (CLIA), 42 U.S.C.A. § 263a(g)(1) (1998).

40. *Final Report* at 126.

41. Ibid.

42. "HCFA to Complete First Round of POL Inspections by May 1995," *BNA Health Care Daily*, October 7, 1994.

43. Ibid. On May 4, 2000, the Department of Health and Human Services proposed new rules ro create a CLIA speciality dealing with genes. 65 *Fed. Register* 25928–25934 (May 4, 2000).

44. *Final Report* at 115.

45. Ibid.

46. Lisa Seachrist, "Testing Genes: Physicians Wrestle with the Information That Genetic Tests Provide," 148, no. 24 *Science News* 394 (December 9, 1995).

47. *Final Report* at xi. ELSI invited organizations interested in genetic testing to nominate potential members of the task force, from which the ultimate fifteen voting members were chosen. Additionally, five agencies in the Department of Health and Human Services (HHS) were invited to send, and sent, nonvoting liaison members to join the task force.

48. Ibid. In the report, the task force did not "recommend policies for specific tests" but instead suggested "a framework for ensuring that new tests meet criteria for safety and effectiveness before they are unconditionally released, thereby reducing the likelihood of premature clinical use" (xii). The task force defined "safety and effectiveness" to include "the validity and utility of genetic tests," as well as laboratory quality assurance, and appropriate usage by both providers and consumers (xi). In particular, the Task Force focused on the use of predictive genetic testing by healthy, or apparently healthy persons—or, put another way, "genetic screening" (xi–xii). The goal of the task force was to recommend policies that would reduce any harm resulting from genetic screening or testing without quashing the benefits (xii).

49. Ibid. at xiv–xv.

50. Ibid. at xv.

51. Ibid.

52. Ibid. at xvi.

53. The task force stated that "members of the [advisory] committee should represent the stakeholders in genetic testing, including professional societies (general medicine, genetics, pathology, genetic counseling), the biotechnology industry, consumers, and insurers, as well as other interested parties. The various HHS agencies with activities related to the development and delivery of genetic tests should send nonvoting representatives to the advisory committee, which can also coordinate the relevant activities of these agencies and private organizations." Ibid. at xvi.

54. Ibid. at xii.
55. See Notice of Establishment of the Secretary's Advisory Committee on Genetic Testing, 63 Fed. Reg. 35242–01 (June 29, 1998). Committee members were appointed a year later. See Press Release, "Shalala Appoints Chair and Members of Genetic Testing Advisory Committee," June 4, 1999.
56. Evelyn Fox Keller, "Nature, Nurture, and the Human Genome Project," 281–299, 292 in Daniel J. Kevles and Leroy Hood, eds., *The Code of Codes: Scientific and Social Issues in the Human Genome Project* (Cambridge: Harvard University Press, 1992).
57. Stuart H. Orkin and Arno G. Motulsky, *Report and Recommendations of the Panel to Assess the NIH Investment in Research on Gene Therapy* (December 7, 1995). http://www.nih.gov/news/panelrep.html.
58. J. G. Wetmur, A. H. Kaya, M. Plewinska, and R. J. Desnick, "Molecular Characterizations of the Human ∂-aminolevulinate Dehydratase 2 (ALAD^2) Allele: Implications for Molecular Screening of Individuals for Genetic Susceptibility to Lead Poisoning," 49 *Am. J. Hum. Genet.* 757–763 (1991).
59. Abby Lippman, "The Genetic Construction of Prenatal Testing: Choice, Consent, or Conformity for Women?" in Karen H. Rothenberg and Elizabeth Thomson, eds., *Women and Prenatal Testing: Facing the Challenges of Genetic Technology* 9–34, 27 (Columbus: Ohio State University Press, 1994).
60. Benno Müller-Hill, "The Shadow of Genetic Injustice," 362 *Nature* 491, 492 (1993). Müller-Hill additionally points out: "Although the molecular-genetic approach will certainly lead to a frenzy of new drugs on the market, in the end the suffering of patients will be helped only partially. . . . It is so much easier to prescribe a pill than to change the social conditions that may be responsible for the severity of symptoms." Ibid.
61. P. N. Goodfellow, "Steady Steps Lead to the Gene," 341 *Nature* 102–103 (1980); G. Vassart, P. Cochaux, M. Abramowicz, "CF Screening," 348 *Nature* 586 (1990); M. Super, M. J. Schwarz, I. B. Sardhawalla, "CF Screening," 344 *Nature* 113–114 (1990); D. Brock, "Population Screening for Cystic Fibrosis," 47 *Am. J. Hum. Genet.* 164–165 (1990).
62. See Barton Childs, "The Clinical Detection of the Genetic Carriers of Inherited Disease," 71 *Medicine* 102 (March 1992), and C. J. Houtchens, "The Humanity in Human Genetics," *USA Weekend*, January 9, 1994, at 20.
63. "Estimates of arrival times for therapeutic benefits run, optimistically, as long as fifty years hence. Thus 'treatment' is at best a long-term goal, and 'prevention' means preventing the births of individuals diagnosed as genetic aberrant—in a word, it means abortion." Keller, "Nature, Nurture, and the Human Genome Project," at 295–296.
64. J. Craig Venter, "Overview and Introduction," (paper presented at the Conference on Mammalian Cloning, Washington, D.C., June 26, 1997).

65. Ibid.

66. Philip Elmer-Dewitt, "The Genetic Revolution," *Time*, January 17, 1994, at 46.

67. Richard Saltus, "4-Year-Old Gets Historic Gene Implant," *Boston Globe*, September 15, 1990, at 1; Barbara J. Culliton, "Gene Therapy Begins," 249 *Science* 1372 (1990).

68. Elmer-Dewitt, "The Genetic Revolution," at 50.

69. Dr. Bernadine Healy, Director, NIH, "Testimony at Hearing on the Possible Uses and Misuse of Genetic Information Before the Subcommittee on Government Information, Justice and Agriculture," 3 *Human Gene Therapy* 51–56, 52 (1991).

70. Michael Conneally (presentation at Symposium on Biology and Culture, Galveston, Texas, November 1996).

71. Laurie Garrett, "The Dots Are Almost Connected . . . Then What?" *Los Angeles Times Magazine*, March 3, 1996, at 22. The 1997 report of the Task Force on Genetic Testing stated that more than one thousand individuals were currently undergoing gene therapy. *Final Report*.

72. See http://www.ncgr.org/gpi.

73. Tim Beardsley, "Vital Data," *Scientific American* 100–105, 105 (March 1996).

74. Ellen Wright Clayton notes: "Patients' experiences almost never get into print." Ellen Wright Clayton, "Screening and Treatment of Newborns," 29 *Hous. L. Rev.* 85, 90 (1992).

75. For some women, the fact that their physician has recommended testing will be enough to make them undergo it. Abby Lippman observes: "Since an expert usually offers testing and careseekers are habituated to follow through with tests ordered by physicians, it is hardly surprising that they will perceive a need to be tested." Abby Lippman, "Prenatal Genetic Testing and Screening: Constructing Needs and Reinforcing Inequities," 18 *Am. J. Law and Med.* 15, 28 (1991). Some couples are compelled to use a prenatal test merely because it exists. S. Adam, S. Wiggins, P. Whyte, M. Bloch, M. H. K. Shokeir, H. Solton, W. Meschino, A. Summers, O. Suchowersky, J. P. Welch, J. Theilmann, and M. R. Hayden, "Five Year Study of Prenatal Testing for Huntington's Disease: Demand, Attitudes, and Psychological Assessment," 30, no. 7 *J. Med. Genet.* 549–556, 549 (1993).

76. Olufunmilayo I. Olopade, "Genetics in Clinical Cancer Care—The Future Is Now," 335 *N. Eng. J. Med.* 1455–1456, 1455 (1996).

77. Statement of Dr. Funmi Olopade, in Peter Gorner, "Cancer Test Ready, But Are We?" *Chicago Tribune*, March 20, 1996, at sec. 1, p. 1.

78. Jeffrey P. Struewing, Dvorah Abeliovich, Tamar Peretz, Naaman Avishai, Michael M. Kaback, Francis S. Collins, and Lawrence C. Brody, "The Carrier Frequency of the BRCA1 185delAG Mutation Is Approximately 1 Percent in Ashkenazi Jewish Individuals," 11 *Nature Genet.* 198–200 (October 1995).

79. See Michael McCarthy, "U.S. Lab Starts Breast-Cancer-Gene Screening," 347 *Lancet* 1033 (1996).

80. Jeffrey P. Struewing, Patricia Hartge, Sholom Wacholder, Sonya M. Baker, Martha Berlin, Mary McAdams, Michelle M. Timmerman, Lawrence C. Brody, and Margaret A. Tucker, "The Risk of Cancer Associated with Specific Mutations of BRCA1 and BRCA2 Among Ashkenazi Jews," 336 *N. Eng. J. Med.* 1401–1408 (1997).

81. Francis S. Collins, "BRCA1—Lots of Mutations, Lots of Dilemmas," 334 *N. Eng. J. Med.* 186–188, 187 (1996).

82. UPI, "Women with Cancer Genes Can Reduce Risk," April 24, 1996. Prophylactic mastectomy had been advocated for some women in high-risk families, even before the discovery of the breast cancer gene. H. T. Lynch, T. Conway, P. Watson, J. Schreiman, and R. J. Fitzgibbons Jr., "Extremely Early Onset Hereditary Breast Cancer (HB): Surveillance/Management Implications," *Nebr. Med. J.* 97–100 (April 1998); M. C. King, S. Rowell, and S. M. Love, "Inherited Breast and Ovarian Cancer: What Are the Risks? What Are the Choices?" 269 *J.A.M.A.* 1975–1980 (1993).

83. Three of twenty-eight women who had oophorectomies in one study had intra-abdominal carcinomatosis. J. K. Tobacman, M. H. Greene, M. A. Tucker, J. Costa, R. Kase, J. F. Fraumeni, "Intra-abdominal Carcinomatosis After Prophylactic Oophorectomy in Ovarian-Cancer-Prone Families," 2 *Lancet* 795–797 (1991).

84. Collins, "BRCA1—Lots of Mutations, Lots of Dilemmas" at 187.

85. Natalie Angier, "Quest to Conquer Breast Cancer," *Los Angeles Daily News*, July 17, 1994.

86. Maureen Henderson, "Current Approaches to Breast Cancer Prevention: Special Report: Breast Cancer Research," 259 *Science* 630 (1993).

87. R. Alta Charo and Karen H. Rothenberg, "'The Good Mother': The Limits of Reproductive Accountability and Genetic Choice," in Rothenberg and Thomson, eds., *Women and Prenatal Testing* 105–130, 109.

88. P. T. Rowley, S. Loader, C. J. Sutera, and M. Walden, "Do Pregnant Women Benefit from Hemoglobinopathy Carrier Detection?" 565 *Ann. N.Y. Acad. Sci.* 152–160 (1989); Lori Andrews, "Genetics and Informed Consent," 271 *Science* 1346–1347 (March 8, 1996) (letter).

89. Calif. Code of Reg. Title 17, § 6527 (1996).

90. Nancy Press and Carol Browner, "Collective Fictions: Similarities in the Reasons for Accepting MSAFP Screening Among Women of Diverse Ethnic and Social Class Backgrounds," 8 *Fetal Diagnosis and Therapy* 97–106 (1993). The authors also note, as does M. Malinowski, "Coming Into Being: Law, Ethics, and the Practice of Prenatal Screening," 45 *Hastings L.J.* 1435–1526, 1493 (1994), that health care professionals may push women into prenatal tests because of fear of malpractice liability.

91. Press and Browner, "Collective Fictions" at 97–106.

92. Dorothy C. Wertz, "Provider's Gender and Moral Reasoning," 8 *Fetal Diagnosis and Therapy* 81, 82 (1993).

93. If a physician or genetic counselor expresses an opinion as to a course of action, a woman may view that opinion as a medical recommendation and may be reluctant to choose an alternative course that the counselor is not "recommending." Elena A. Gates, "The Impact of Prenatal Testing on Quality of Life in Women," 8 *Fetal Diagnosis and Therapy* 236–243, 240 (supp. 1) (1993).

94. Genetic counselors point out that physicians and counselors will "get directive, especially if they feel the diagnosis is extremely severe or extremely mild." Malinowski, "Coming Into Being" at 1468. According to Judy Norsigian of the Boston Women's Health Book Collective, "When it comes to something like Down's syndrome, most physicians have been extremely directive and even obnoxious. They will even say, 'we'll be scheduling an abortion for you.' This happens even when the extent of the disability is very mild." Charlotte Allen, "Boys Only; Pennsylvania's Anti-Abortion Law," 206 *New Republic* 16 (1992).

95. Bonnie Gangelhoff, "Tragedy Spurs Mother to Back Registry," *Houston Post*, July 1993, A1, A20 at A17 (quoted in Malinowski, "Coming Into Being" at 1522).

96. Theresa Marteau, Harriet Drake, Margaret Reid, Maria Feijoo, Maria Soares, Irma Nippert, Peter Nippert, and Martin Bobrow, "Counselling Following Diagnosis of Fetal Abnormality: A Comparison Between German, Portuguese, and U.K. Geneticists," 2 *Eur. J. Hum. Genet.* 96–102, 99 (1994).

97. Theresa M. Marteau and Harriet Drake, "Attribution for Disability: The Influence of Genetic Screening," 40 *Soc. Sci. & Med.* 1127, 1129 (1995).

98. Ibid. at 1130.

99. Peter E. S. Freund and Meredith B. McGuire, *Health, Illness, and the Social Body: A Critical Sociology* 240, 243–44 (Englewood Cliffs, N.J.: Prentice Hall, 1991).

100. "CDC Speaks Out on Gene Testing: Won't Scan Healthy Populations," *Medical Utilization Management*, May 28, 1998 (reported by Brian Ward, vice president of laboratory operations of Myriad Genetics, which provides the breast cancer test).

101. Gail Geller, Ellen S. Tambor, Gary A. Chase, and Neil A. Holtzman, "Measuring Physicians' Tolerance for Ambiguity and Its Relationship to Their Repeated Practices Regarding Genetic Testing," 31 *Medical Care* 989–1001 (1993).

102. Sandra Leiblum and Christopher Barbrack, "Artificial Insemination by Donor: A Survey of Attitudes and Knowledge in Medical Students and Infertile Couples," 15 *J. Biosoc. Sci.* 165, 170 (1983).

103. *Final Report* at 65.

104. See Andrews et al., *Assessing Genetic Risks.*

105. Ellen Wright Clayton, "CF Pilot Study" (presentation at Cystic Fibrosis Grantee Meeting, National Institutes of Health, Bethesda, Maryland, September 8–10, 1993).

106. M. M. McGovern, J. D. Goldberg, and K. J. Disnich, "Acceptability of Chorionic Villi Sampling for Prenatal Diagnosis," 155 *Am. J. Obstet. Gyn.* 25 (1986).

107. Lippman, "The Genetic Construction of Prenatal Testing" at 31 n. 2

108. Neil A. Holtzman (presentation at "Genetic Testing for Breast Cancer Susceptibility: The Science, the Ethics, the Future," symposium held in San Francisco, California, November 22, 1996).

109. R. F. Weir and J. R. Horton, "DNA Banking and Informed Consent—Part 1," 17 *IRB* 1–4, 4 (July to August 1995).

110. R. F. Weir and J. R. Horton, "DNA Banking and Informed Consent—Part 2," 17 *IRB* 1–8, 5–6 (September–December 1995).

111. J. E. McEwen and P. R. Reilly, "A Review of State Legislation on DNA Forensic Data Banking," 54 *Am. J. Hum. Genet.* 941–958 (1994); W. McGoodwin, "Genie Out of the Bottle: Genetic Testing and the Discrimination It's Creating," *Washington Post,* May 5, 1996, at G3.

112. McEwen and Reilly, "Survey of DNA Diagnostic Laboratories Regarding DNA Banking," 56 *Am. J. Hum. Genet.* 1477–1486 (1995).

113. M. K. Cho, M. Arruda, and N. A. Holtzman, "Appendix 4: Informational Materials About Genetic Tests," in *Final Report* at 125–135, 125 (September 1997). The authors identified 178 organizations, including biotechnology companies, molecular genetics laboratories, and cytogenetics laboratories, that offered printed or audiovisual information about their services. Of the 178 organizations, 169 (95 percent) responded to the authors' survey.

114. Ibid. at 127.

115. Ibid. at 135 (table 2).

116. *Final Report* at 56.

117. "Public Education in Genetics," in Andrews et al., *Assessing Genetic Risks,* 185–201.

118. National Bioethics Advisory Commission, *Cloning Human Beings: Report and Recommendations of the National Bioethics Advisory Commission* v (Rockville, Md., June 1997).

119. "News from the Rockefeller University, 'Fat, Body Weight Regulated by Newly Discovered Hormone'" (press release, July 27, 1995).

120. Robert V. Considine, Madhur K. Sinha, Mark L. Heiman, Aidus Kriauciunas, Thomas W. Stephens, Mark R. Nyce, Joanna P. Ohannesian, Cheryl C. Marco, Linda J. McKee, Thomas L. Bauer, and José F. Caro, "Serum Immunoreactive-Leptin Concentrations in Normal-Weight and Obese Humans," 334 *N. Eng. J. Med.* 292–295 (1996).

121. Stuart H. Orkin and Arno Motulsky, co-chairs, *Report and Recommendations of the Panel to Assess the NIH Investment in Research on Gene Therapy* 1, 3 (December 7, 1995). By early 2000, more than 4,000 people had participated in nearly 400 gene-therapy experiments. Marlene Cimons, "NIH to Order New Reports on Pst Gene Therapy Cases," *Los Angeles Times*, February 28, 2000 at 1A.

122. Orkin and Motulsky, *Report and Recommendations*, at 2.

123. Neil A. Holtzman (presentation analyzing brochures of companies).

124. Mark A. Kay, Catherine S. Manno, Margaret V. Ragni, Peter J. Larson, Linda B. Couto, Alan McClelland, Bertil Glader, Amy J. Chew, Shing J. Tai, Roland W. Herzog, Valder Arruda, Fred Johnson, Ciaran Scallan, Erik Skarsgard, Alan W. Flake, and Katherine A. High, "Evidence for Gene Transfer and Expression of Factor IX in Haemophlia B Patients Treated with an AAV Vector," 24 *Nature Genet.* 257–261 (2000).

125. Rick Weiss, "Cautions Over Gene Therapy Put Hopes on Hold," *Washington Post*, March 7, 2000, at A01.

126. Ibid.

127. Marina Cavazzana-Calvo, Salima Hacein-Bey, Genevieve de Saint Basile, Fabian Gross, Eric Yvon, Patrick Nusbaum, Francoise Selz, Christophe Hue, Stephanie Certain, Jean-Laurent Casanova, Philippe Bousso, Francoise Le Deist, Alain Fischer, "Gene Therapy of Human Severe Combined Immunodeficiency (SCID)-X1 Disease," 288 *Science* 669–672, (April 28, 2000).

128. "Medical Researchers in France Have Successfully Treated Some Children to Rid Them of Immune Deficiency Disease by Using Genetic Tranplantation Therapy," *All Things Considered*, National Public Radio, April 27, 2000.

129. Similarly, when the director of the National Institutes of Health at the time, Bernadine Healy, testified in Congress, she described the first gene therapy experiment as a success, failing to point out that the children were also receiving standard medical treatments for the disorder, thus making it impossible to determine if the gene therapy made a difference. Bernadine Healy, "Testimony at Hearing on the Possible Uses and Misuses of Genetic Information, Before the Subcommittee on Government Information, Justice and Agriculture, 3 *Human Gene Therapy* 51–56 (October 17, 1991).

130. In his second inaugural address, President Clinton said: "Scientists are now decoding the blueprint of human life. Cures for our most feared illnesses seem close at hand." *Baltimore Sun*, January 21, 1997, at 8A.

131. This distortion has an impact on patients' treatment choices. Some cystic fibrosis carriers choose not to undergo amniocentesis because they believe gene therapy for cystic fibrosis will be available shortly.

132. Dorothy Nelkin and Richard Nelson, with Casey Kiernan, "Commentary: University-Industry Alliances," 12 *Science, Technology, and Human Values* 65–74, 72 (1987).

133. Ibid.

134. Andrews et al., *Assessing Genetic Risks* at 195.

135. *Report of the Working Group of the Stanford Program in Genomics, Ethics and Society on Genetic Testing for Breast Cancer Susceptibility* at 12 (November 11, 1996).

7. Impact of Genetics on Cultural Value and Social Institutions

1.   Frederick Hecht and Barbara Kaiser McKaw, "Chromosome Instability Syndromes," in John J. Mulvihill, Robert W. Miller, and Joseph F. Fraumeni Jr., *Genetics of Human Cancer* 105, 114 (New York: Raven Press, 1977).

2.   Fergus Clydesdale, "Present and Future of Food Science and Technology in Industrialized Countries," *Food Technology* 134–136 (September 1989), cited in Dorothy Nelkin, "The Social Dynamics of Genetic Testing: The Case of Fragile-X," 10, no. 4 *Medical Anthropology Quarterly* 537–550, 546 (1996).

3.   See, e.g., Margery Shaw, "Conditional Prospective Rights of the Fetus," 5 *J. Legal Med.* 63 (1984).

4.   Dorothy Nelkin, "Genetics and Dismantling the Welfare State," in Ronald A. Carson and Mark A. Rothstein, eds., *Behavioral Genetics: The Clash of Culture and Biology* 156–171, 165 (Baltimore: Johns Hopkins University Press, 1999).

5.   Criticizing such an approach, geneticist Benno Müller-Hill points out: "The scientific prediction of a person's limitations, and thus his possible fate, has a very dangerous component in that it may lead an individual to inaction and despair. It may also lead the population to believe that, as there is no real chance, money should not be wasted to counteract genetic limitations. It could be forgotten that these limitations are also set by environmental factors." Benno Müller-Hill, "The Shadow of Genetic Injustice," 362 *Nature* 491, 492 (April 8, 1993).

6.   Jeremy Webb, "A Fragile Case for Screening? Evaluating the Efficacy of Screening for Fragile X Syndrome," 140 *New Scientist* 10 (December 25, 1993); Erik Lineala, "Renewed Debate Surfaces Around Human Genome Project," 20, no. 4 *Alternatives* (September 1994); Diane E. Lewis, "Under a Genetic Cloud," *Boston Globe*, August 14, 1994, at A1.

7.   Webb, "A Fragile Case for Screening?"

8.   *National Commission for Control of Huntington's Disease and Its Consequences, Report Volume I: Overview* 85, publication no. NIH 78–1501 (Washington, D.C.: U.S. Department of Health, Education, and Welfare, 1977).

9.   Humphrey Taylor, "Harris Poll," *Gannett News Service*, May 28, 1995.

10. For information on how insurers have responded to information about people's genetic status, see Paul R. Billings, Mel A. Kohn, Margaret de Cuevas, Jonathan

Beckwith, Joseph S. Alper, and Marvin R. Natowicz, "Discrimination as a Consequence of Genetic Testing," 50 *Am. J. Hum. Genet.* 476–482 (1992); Office of Technology Assessment, U.S. Congress, *Cystic Fibrosis and DNA Tests: Implications of Carrier Screening* (Washington, D.C.: U.S. Government Printing Office, 1992). See also Neil A. Holtzman, *Proceed with Caution* 195 (Baltimore: Johns Hopkins University Press, 1989), and Larry Gostin, "Genetic Discrimination: The Use of Genetically Based Diagnostic and Prognostic Tests by Employers and Insurers," 17 *Am. J. Law and Med.* 109–144, 119 (1991).

11. E. Virginia Lapham, Chahira Kozma, and Joan O. Weiss, "Genetic Discrimination: Perspectives of Consumers," 274 *Science* 621–624, 622 (1996). In another study, 14 percent of responding genetic counselors reported that their counselees had difficulty obtaining or retaining health insurance because of genetic testing results. Lori B. Andrews, Jane E. Fullarton, Neil A. Holtzman, and Arno G. Motulsky, eds., *Assessing Genetic Risks: Implications for Health and Social Policy*, 270 (Washington, D.C.: National Academy Press, 1994).

12. Stephanie Armour, "Could Your Genes Hold You Back?" *USA Today*, May 5, 1999, at 1B.

13. See Office of Technology Assessment, *Cystic Fibrosis and DNA Tests* at 33.

14. National Action Plan on Breast Cancer and NIH-DOE Working Group on Ethical, Legal, and Social Implications of Human Genome Research, Conference on Genetic Discrimination and Health Insurance: A Case Study on Breast Cancer, Testimony of Mary Jo Ellis Kahn.

15. Billings et al., "Discrimination as a Consequence of Genetic Testing" at 478.

16. There is some limitation for doing this under group health insurance plans because of the federal Health Insurance Portability and Accountability Act, discussed *infra*.

17. "CDC Speaks Out on Gene Testing: Won't Scan Healthy Population," *Medical Utilization Management*, May 28, 1998.

18. Nancy Axelrod Comer (presentation at Breast Cancer Briefing, Susan G. Komen Breast Cancer Foundation, May 14, 1996). (Aetna Insurance Company asks if the individual has had genetic testing.)

19. Caryn Lerman, Janet Seay, Andrew Balshem, and Janet Audrain, "Interest in Genetic Testing Among First-Degree Relatives of Breast Cancer Patients." 57 *Am. J. Med. Genet.* 385–392, 389 (1995).

20. Testimony of Caryn E. Lerman, Senate Cancer Coalition, September 29, 1995.

21. Bob Groves, "New Privacy Fight Is All in the Genes," *Bergen Record*, July 18, 1999, at N04.

22. Wendy McGoodwin, Council for Responsible Genetics (presentation at ABA Annual Meeting, August 1997).

23. Tim Beardsley, "Vital Data," *Scientific American* 100–105, 102 (March 1996).

24. Ibid. at 103.

25. Ibid. (interview with William C. Dickson, research management chair of the VHL Family Alliance).
26. Ibid. (interview with Gregory G. Germino, a researcher at Johns Hopkins University School of Medicine).
27. Armour, "Could Your Genes Hold You Back?"
28. A community rating system "effectively spreads the risks of health care utilization over a substantial portion of the population which creates a more equitable distribution of the financial burden." Katy Chi-Wen Li, "The Private Insurance Industry's Tactics Against Suspected Homosexuals: Redlining Based on Occupation, Residence, and Marital Status," 22 *Am. J. Law and Med.* 477 (1996) at 500. At least nine states have enacted laws that require community rating in some instances of health insurance. Ark. Stat. Ann. § 23–98–110; Fla. Stat. ch. 627.6699; Ky. Rev. Stat. § 304.17A-120; La. Rev. Stat. § 22:228.6; 1997 N.H. Laws 344; N.Y. Comp. Codes R. and Regs. tit. 11, § 360.1-.12; N.C. Gen. Stat. § 58–50–130; Vt. Stat. Ann. § 4080b; Wash. Rev. Code Ann. § 48.43.005 et seq.
29. Neil A. Holtzman and David Shapiro, "Genetics and Public Policy," 316 *Brit. Med. J.* 852 (1998).
30. See, e.g., N.J. Stat. Ann. § 17B:30–12; Or. Rev. Stat. § 746.135.
31. Lori Andrews, *Medical Genetics: A Legal Frontier* 18 (Chicago: American Bar Foundation, 1987).
32. Billings et al., "Discrimination as a Consequence of Genetic Testing" at 478.
33. Jon Matthews, "Bias Based on Genetic Testing Techniques," *Sacramento Bee,* May 7, 1995.
34. Groves, "New Privacy Fight Is All in the Genes."
35. "Information About Klinefelter Syndrome," at http://www.genetic.org/ks/aboutxxy.htm.
36. Groves, "New Privacy Fight Is All in the Genes."
37. Ibid.
38. Gina Kolata, "Tests to Assess Risks for Cancer Raising Questions," *New York Times,* March 27, 1995, at A1.
39. Dorothy C. Wertz and John C. Fletcher, *Ethics and Human Genetics: A Cross-Cultural Perspective* (Berlin, N.Y.: Springer-Verlag, 1989).
40. Dorothy Nelkin and Lawrence Tancredi, *Dangerous Diagnostics: The Social Power of Biological Information* 164 (New York: Basic Books, 1989).
41. Survey quoted in "Genetic Discrimination: The Next Civil Rights Issue," *Health Line,* May 6, 1999. See also Armour, "Could Your Genes Hold You Back?"
42. 1999 AMA Survey on Workplace Testing: Medical Testing: Summary of Key Findings at 2. http://www.amanet.org/research/medico/index.htm. An earlier study, a 1989 survey of companies by the Office of Technology Assessment,

found that one in twenty companies conducted genetic screening or monitoring in the workplace. Office of Technology Assessment, U.S. Congress, *Genetic Monitoring and Screening in the Workplace* 22 (1990).

43. Armour, "Could Your Genes Hold You Back?"
44. *Norman-Bloodsaw v. Lawrence Berkeley Lab*, No. C-95–03220-VRW (N.D. Calif. filed September 12, 1995).
45. "Case Law," *Employment Testing—Law and Policy Reporter* 138 (September 1996), discussing *Norman-Bloodsaw v. Lawrence Berkeley Laboratory*, No. C-95–3220-VRW (N.D. Cal., June 10, 1996). The judge also indicated that the intrusion was not actionable, since the workers' employment apparently had not been affected by the test. Sally Lehrman, "Berkeley Employees Fight Ruling on Gene Tests Without Consent," *Biotechnology Newswatch*, May 5, 1997.
46. See *Norman-Bloodsaw v. Lawrence Berkeley Laboratory*, 135 F.3d 1260, 1268 (9th Cir. 1998).
47. Ibid. at 1269.
48. Ibid. at 1275.
49. Ibid. at 1269.
50. Ibid.
51. Ibid. at 1270.
52. Ibid. at 1272.
53. Ibid.
54. Ibid.
55. See 972 P.2d 1060 (Colo. Ct. App. 1998).
56. Billings et al., "Discrimination as a Consequence of Genetic Testing"; Office of Technology Assessment, *Cystic Fibrosis and DNA Tests*; Holtzman, *Proceed with Caution*; Gostin, "Genetic Discrimination."
57. Karen H. Rothenberg, "Genetic Information and Health Insurance: State Legislative Approaches," 23 *J. Law, Med. and Ethics* 312–319 (1995).
58. Del. Code Ann. tit 16, § 1221(a); Fla. Stat. ch. 760.40; Nev. Rev. Stat. § 629.151; N.M. Stat. Ann. § 24–21–4; N.Y. Civ. Rights § 79–I; Vt. Stat. Ann. tit. 18, § 9332.
59. Del. Code Ann. tit 16, § 1221(b)(1); Fla. Stat. ch. 760.40(2)(a); Nev. Rev. Stat. § 629.151(1); N.M. Stat. Ann. § 24–21–3(c)(1); Vt. Stat. Ann. tit. 18, § 9332(3).
60. Del. Code Ann. tit 16, § 1221(b)(5); Nev. Rev. Stat. § 629.151(4); N.M. Stat. Ann. § 24–21–4(c)(9); Vt. Stat. Ann. tit. 18, § 9332(5)(d).
61. Twenty-nine states prohibit denial of insurance based on certain genetic information. See Ala. Code § 27–53–2(b); Ariz. Rev. Stat. Ann. § 20–448(D); Cal. Ins. Code § 10140(b); Colo. Rev. Stat. Ann. § 10–3–1104.7(D); Conn. Gen. Stat. § 38a–816(19); Del. Code Ann. tit. 18, 2317(B); Fla. Stat. Ann. § 627.4301(2)(a); Ga. Code Ann. § 33–51–1(4); Haw. Rev. Stat. § 431:10A-118(a)(1); Ind. Code § 27–8–26–7; Kan. Stat. Ann. § 40–2259(b)(3); Ky. Rev. Stat. Ann. § 304.12–085(2)(a); La. Rev. Stat. Ann. § 22:213.7(b)(1)(a); Md.

Code Ann., Ins. § 27–909(c)(1); Minn. Stat. § 72A-139(3); Mont. Code Ann. § 33–18–206(3); N.H. Rev. Stat. Ann. § 141-H:2(II); N.J. Stat. Ann. § 17B:30–12(e); N.M. Stat. Ann. § 24–21–4(b); N.C. Gen. Stat. § 58–3–215(c)(2); Ohio Rev. Code Ann. § 1751.64(b)(4); Okla. Stat. tit. 36, § 3614.1(C)(2); Or. Rev. Stat. § 746.135(3); S.C. Code Ann. § 38–93–20(1); Tenn. Code Ann. § 56–7–2703; Tex. Ins. Code Ann. § 21.73 Sec. 3(a); Vt. Stat. Ann. tit. 18, § 9334; Va. Code Ann. Sec. 38.2–508.4(b)(1); and Wis. Stat. § 631.89(2)(d). Four states prohibit insurers from requiring genetic tests. See 410 Ill. Comp. Stat. 513/20(a); Me. Rev. Stat. Ann. tit. 24-a, § 2159(c)(2); Mo. Rev. Stat. § 375.1303; and Nev. Rev. Stat. § 695B.317. In addition, Nebraska has established a commission to review genetic discrimination in insurance. Neb. Code Ann. § 71–8102.

62.   See Ala. Code § 27–53–2(b); Ariz. Rev. Stat. Ann. § 20–448(D); Cal. Ins. Code § 10140(b); Colo. Rev. Stat. § 10–3–1104.7(D); Conn. Gen. Stat. § 38a–816(19); Del. Code Ann. tit. § 18 2317(b); Fla. Stat. ch. 627.4301(2)(a); Ga. Code Ann. § 33–51–1(4); Haw. Rev. Stat. § 431:10A-118(a)(1); Ind. Code § 27–8–26–7; Kan. Stat. Ann. § 40–2259(b)(3); Ky. Rev. Stat. Ann. § 304.12–085(2)(a); La. Rev. Stat. Ann. § 22:213.7(b)(1)(a); Md. Code Ann., Ins. § 27–909(c)(1); Minn. Stat. § 72A-139(3); Mont. Code Ann. § 33–18–206(3); N.H. Rev. Stat. Ann. § 141-H:2(II); N.J. Rev. Stat. § 17B:30–12(e); N.M. Stat. Ann. § 24–21–4(b); N.C. Gen. Stat. § 58–3–215(c)(2); Ohio Rev. Code Ann. § 1751.64(b)(4); Okla. Stat. tit. 36 § 3614(C)(2); Or. Rev. Stat. § 746.135(3); S.C. Code Ann. § 38–93–20(1); Tenn. Code Ann. § 56–7–2703; Tex. Ins. Code Ann. § 21.73 Sec. 3(a); Vt. Stat. Ann. tit. 18 § 9334; Va. Code Ann. Sec. 38.2 § 508.4(b)(1); and Wis. Stat. § 631.89(2)(d).

63.   See Ala. Code § 27–53–2(b); Ariz. Rev. Stat. Ann. § 20–448(F); Cal. Ins. Code § 10143; Conn. Gen. Stat. § 38a–816(19); Haw. Rev. Stat. § 431:10A-118(a)(1); Ind. Code § 27–8–26–5(3); Kan. Stat. Ann. § 40–2259(b)(4); Ky. Rev. Stat. Ann. § 304.12–085(1); La. Rev. Stat. Ann. § 22:213.7(b)(1); Me. Rev. Stat. Ann. tit. 24, § 2159(c)(2); Md. Code Ann., Ins. § 27–909(c)(1); Minn. Stat. § 72A.139(3)(4); Mo. Rev. Stat. § 375.1303; Mont. Code Ann. § 33–8–206(14); Nev. Rev. Stat. § 689(A)417(c); N.H. Rev. Stat. Ann. § 141-H:4(IV); N.M. Stat. Ann. § 24–21–4(b); N.C. Gen. Stat. § 58–3–215(c)(1); Ohio Rev. Code Ann. § 1751.64(b)(2); Okla. Stat. tit. 36, § 3614.1(C)(2); Or. Rev. Stat. § 746.135(3); S.C. Code Ann. § 38–93–20(1); Tenn. Code Ann. § 56–7–2703; Tex. Ins. Code Ann. § 21.73 Sec. 3(a); Va. Code Ann. Sec. 38.2-508.4(b)(6); and Wis. Stat. § 631.89(2)(d).

64.   This is because only nine states prohibit considering information from a relative in determining rates. See Haw. Rev. Stat. § 431:10A-118(a)(1); Kan. Stat. Ann. § 40–2259(b)(4); La. Rev. Stat. Ann. § 22:213.7(B)(1); Minn. Stat. § 72A.139(3)(4); N.J. Rev. Stat. § 17B:30-12(e); Nev. Rev. Stat. § 689A.417; N.H. Rev. Stat. Ann. § 141-H:4(IV); Tenn. Code Ann. § 56–7–2703; and Wis. Stat. § 631.89(2)(d).

65. See Fla. Stat. Ann. § 627.4301(2); Ky. Rev. Stat. Ann. § 304.12–085(2)(b); La. Rev. Stat. Ann. § 22:213.7(b)(1); Me. Rev. Stat. Ann. tit. 24, § 2159(c)(2); Minn. Stat. § 72A.139(3)(3); Mo. Rev. Stat. § 595.105(2); Nev. Rev. Stat. Ann. § 689.417(b); N.H. Rev. Stat. Ann. § 141-H:2(II); N.M. Stat. Ann. § 24–21–4(A); Tenn. Code Ann. § 56–7–2703; Va. Code Ann. § 38.2–508.4(b); and Wis. Stat. § 631.89(2)(b).

66. 29 U.S.C. § 1001 *et seq.*, at § 1144(a).

67. At least 65% of all companies and 82% of companies with more than 5,000 employees are self-insured. Eric Zicklin, "More Employers Self-Insure Their Medical Plans Survey Finds," *Business and Health*, April 1992, at 74.

68. 42 U.S.C.S. § 300gg *et seq.*

69. 42 U.S.C.S. § 300gg(b)(1)(B).

70. See Ariz. Rev. Stat. 41–1463(4); Conn. Gen. Stat. § 46a-60(11)(B); Del. Code Ann. tit. 19, 711(a); Iowa Code § 729.62(b); Me. Rev. Stat. Ann. tit. 5, § 19302, Mo. Rev. Stat. § 375.1306(1); N.H. Rev. Stat. Ann. § 141H:3(I)(b); N.J. Stat. Ann. § 10:5–12(a); N.Y. Exec. Law § 296(1); N.C. Gen. Stat. § 95–28.1A; Okla. Stat. tit. 36, § 3614.2(C)(1); Or. Rev. Stat. § 659.036(1); R.I. Gen. Laws § 28–6.7–1(2); Tex. Lab. Code Ann. § 21.402(1); Vt. Stat. Ann. tit. 18, § 9333(a); and Wis. Stat. § 111.372(1)(b).

71. Del. Code Ann. tit. 19, § 711(e); Me. Rev. Stat. tit. 5, § 19302 and tit. 5, § 19301; N.J. Stat. Ann. § 10:5–12(a); N.Y. CLS Exec. Law § 292(21)(d) (prohibits discrimination based on offspring's test); N.C. Gen. Stat. § 95–28.1A(a) and (b).

72. 42 U.S.C.A. § 12101–12117.

73. 1990 U.S. Dist Lexis 4070 (1990). This was a case brought before the passage of the ADA. It alleged racial discrimination, but the court held that a discrimination claim was inappropriate because the person's genetic disorder, sickle-cell anemia, rendered him unqualified for the job. Interestingly, even though the plaintiff had apparently misrepresented his condition on his application to be a firefighter (saying he had sickle-cell carrier status, rather than sickle-cell anemia), he was able to receive lifelong disability retirement benefits from the Kansas Police and Fire Pension System.

74. U.S. Equal Employment Opportunity Commission, *Compliance Manual*, § 902, "Definition of the Term 'Disability'" 47 (1995).

75. Ibid.

76. David L. Coleman, "Who's Guarding Medical Privacy?" 17 *Business and Health* 29 (March 1, 1999).

77. Mark Siegler, "Confidentiality in Medicine—A Decrepit Concept," 307 *N. Eng. J. Med.* 1518, 1519 (1982).

78. Ibid.

79. *McPheeters v. Bd. of Med. Exam'rs*, 103 Cal. App. 297, 284, 284 P. 938 (1930) (chatty letter from physician to his former assistant).

80. Lori Andrews and Ami Jaeger, "Confidentiality of Genetic Information in the Workplace," 17 *Am. J. Law and Med.* 75–108 (1991).

81. Andrews, *Medical Genetics* at 191.

82. *Doe v. Southeastern Pa. Transp. Auth.*, 72 F.3d 1133 (3d Cir. 1995), *cert. denied*, 519 U.S. 808 (1996).

83. Coleman, "Who's Guarding Medical Privacy?"

84. Ibid. This is based on a 1993 Louis Harris Poll.

85. Ibid.

86. Ibid.

87. *Berkeley County Dept. of Social Services v. David Galley and Kimberly Galley*, 92-DR-08–2699 (Moncks Corner, South Carolina, April 19, 1994).

88. Mark Rothstein, "Preventing the Discovery of Plaintiff Genetic Profiles by Defendants Seeking to Limit Damages in Personal Injury Litigation," 71 *Ind. L.J.* 877 (1996).

89. Meg Fletcher, "Genetic Testing Ordered in Product Liability Case," *Business Insurance*, August 1, 1994, at 1.

90. Ruth Hubbard and Elijah Wald, *Exploding the Gene Myth* (Boston: Beacon, 1993). Rebecca Pentz has pointed out that tobacco companies "claim that smoking only causes cancer in those with a genetic susceptibility—implying that if we just fix the genes there is no need to stop buying and using their product." Rebecca D. Pentz, "Commentary on 'Ethical Issues in Genetic Engineering, Screening and Testing,' by P. Michael Conneally," 1–2 (paper prepared for Symposium on Biology and Culture, Galveston, Texas, November 1996).

91. Don Babwin, "Meet the New Building Block of Legal Life," *Illinois Legal Times*, January 1999.

92. Ibid., citing *Dewey v. Zack.*

93. H. G. Brunner, M. Nelen, X. O. Brakefield, H. H. Ropers, and B. A. van Oost, "Abnormal Behavior Associated with a Point Mutation in the Structural Gene for Monoamine Oxidase A," 262 *Science* 578 (October 22, 1993).

94. Contrast *In re Ewaniszyk*, 788 P.2d 690 (Cal. Ct. App. 1990), with *Baker v. State Bar of California*, 781 P.2d 1344 (Cal. 1989).

95. Associated Press, "Disease Cited in Murder Acquittal," *Cleveland Plain Dealer*, September 29, 1994.

96. Sarah Boseley, "Second Front: Genes in the Dock," *Guardian*, March 13, 1995, at T2.

97. Such an approach has allowed the CDC to enhance its budget. President Clinton endorsed a special CDC line item to fund violence-prevention programs. Sheryl Stolberg, "Fear Clouds Search for Genetic Roots of Violence," *Los Angeles Times*, December 31, 1993, at A1 (part 2). John Douard describes how turning a social problem into a public health problem is done as an "in-

stitutionalized distraction." When a society throws up its hands about doing anything about racism or poverty, it can appease itself by moving to a new type of expert, a public health expert. John Douard (presentation at Symposium on Biology and Culture, Galveston, Texas, November 1996).

98. Sheryl Stolberg, "Fear Clouds Search for Genetic Roots of Violence," *Los Angeles Times*, December 30, 1993, at A1 (part 1).

99. Ibid.

100. Brunner et al., "Abnormal Behavior Associated with a Point Mutation in the Structural Gene for Monoamine Oxidase A."

101. Ibid.

102. Charles C. Mann, "Behavioral Genetics in Transition," 264 *Science* 1686–1689, 1689 (1994).

103. Jonathan Beckwith, "Social and Political Uses of Genetics in the United States: Past and Present," 265 *Ann. N.Y. Acad. Sci.* 46, 54 (1976). Currently, society seems to be willing to find medical solutions to criminal acts. In California, on August 31, 1996, the legislature passed a bill permitting chemical castration of child molesters. Calif. Penal Code § 645 (1998). See Dan Morain and Max Vanzi, "Senate OKs Access to Sex Offenders Database," *Los Angeles Times*, September 1, 1996, at A3.

104. Harold P. Green, "Genetic Technology: Law and Policy for the Brave New World," 48 *Ind. L.J.* 559, 571 (1973).

105. Ericka L. Johnson, "'A Menace to Society': The Use of Criminal Profiles and Its Effect on Black Males," 38 *Howard L.J.* 629–664, 636–637 (1995). For example, white pregnant women are slightly more likely to abuse drugs than African American pregnant women, but African American pregnant women are 9.58 times as likely to be reported for substance abuse during pregnancy. Kary Moss, "Substance Abuse During Pregnancy," 13 *Harv. Women's L.J.* 278–299, 294 (1990) (citation omitted). Moreover, offenses that are seen as primarily African American are punished more harshly than those viewed as white offenses — for example, the use of crack cocaine is subject by statute to longer prison sentences than is the use of powder cocaine. Johnson, "A Menace to Society" at 644. More generally, surveillance has been used discriminatorily against men of color, to the point where it has been found justifiable to detain African American men or Hispanic men and search them if they are found in primarily white neighborhoods. Ibid. at 630–631.

106. Sheryl Stolberg, "Fear Clouds Search for Genetic Roots of Violence," *Los Angeles Times*, December 30, 1993, at A1 (part I).

107. Beverly Horsburgh points out: "Intellectuals who participate within a discipline become both subjects and objects, acted upon by the very body of knowledge they create. As part of a hierarchal structure, some scientists are thereby inevitably drawn to eugenics. By its nature, science demands reductionist thinking,

examining only isolated parts of the world or of human beings. Individuals all too easily can be reduced to the purported worth of their genes." Beverly Horsburgh, "Schrodinger's Cat, Eugenics, and Compulsory Sterilization of Welfare Mothers: Deconstructing the Old/New Rhetoric and Constructing the Reproductive Rights of Natality for Low-Income Women of Color," 17 *Cardozo L. Rev.* 531, 534 (1996). Steven Rose puts it this way: "The core issue is reducibility, which as Medavan once remarked, comes not as second but as first nature to natural scientists." Steven Rose, "The Rise of Neurogenetic Determinism," 372 *Nature* 380–382, 390 (1995).

108. Walter Gilbert, "A Vision to the Grail," 83–87 in Leroy Hood and Daniel J. Kevles, eds., *Code of Codes: Scientific and Social Issues in the Human Genome Project* (Cambridge: Harvard University Press, 1992).

109. Tom Wolfe, "Sorry, But Your Soul Just Died," *Forbes ASAP*, December 2, 1996, at 211–223, 212.

110. Leon Jaroff, "The Gene Hunt," *Time*, March 20, 1989, at 63.

111. Barbara Katz Rothman, *Genetic Maps and Human Imaginations: The Limits of Science in Understanding Who We Are* 13 (New York: Norton, 1998).

112. Jeffrey Obser, "Genetic Pressures/Experts are Pushing More Testing for Diseases," *Newsday*, February 24, 1998, at C03.

113. Dorothy Nelkin and M. Susan Lindee, *The DNA Mystique: The Gene as Cultural Icon* 82 (New York: Freeman, 1995).

114. Philip Reilly, "Eugenic Sterilization in the United States," in Aubrey Milunsky and George Annas, eds., *Genetics and the Law III* 227, 231 (New York: Plenum, 1985).

115. Beckwith, "Social and Political Uses of Genetics in the United States" at 48. Sanger wrote: "More children from the fit, less from the unfit—that is the chief issue of birth control." Ibid. See also Linda Gordon, *Woman's Body, Woman's Right: A Social History of Birth Control in America* 281–282 (New York: Grossman, 1976).

116. Daniel Kevles, *In the Name of Eugenics: Genetics and the Uses of Human Heredity* 63 (New York: Knopf, 1985).

117. Ibid. at 61. One rabbi told his congregation, "May we do nothing to permit our blood to be adulterated by infusion of blood of inferior grade." Ibid.

118. Ibid. at 58.

119. For example, Cyril Burt fabricated IQ data in identical twins. Stephen Jay Gould, *The Mismeasure of Man* 27 (New York: Norton, 1981).

120. Like other eugenicists, he believed that morality was linked to intelligence. Ibid. at 160.

121. Ibid.

122. Ibid. at 168.

123. Ibid.

124. Ibid.

125. Ibid. at 169.

126. Ibid. at 171, 173.

127. "Clinton's Inaugural Address," *Baltimore Sun*, January 21, 1997, at 8A.

128. "Results of Public Survey on Human Genetics Released," *Cancer Weekly*, December 21, 1992, at 9. See also Richard Liebmann-Smith, "It's a (Blond-Haired, Blue-Eyed, Even-Tempered, Ivy-Bound) Boy!" *New York Times*, February 7, 1993, at sec. 6, p. 21.

129. Ibid.

130. Charles C. Mann, "Behavioral Genetics in Transition," 264 *Science* 1686–1689, at 1686 (1994). Recent research claims to have found a genetic disposition for pathological gambling. Ted Gregory, "Researcher's Theory Criticized: Doctor Proposes Genetic Link to Gambling," *Chicago Tribune*, September 4, 1996, at sec. 2, p. 8.

131. For a discussion of the numerous failures to replicate researchers' genetic linkage of common neuropsychiatric disorders (such as schizophrenia, manic-depression, and Alzheimer's disease), see Neil Risch, "Genetic Linkage and Complex Diseases, with Special Reference to Psychiatric Disorders," 7 *Genetic Epidemiology* 3–16 (1990). Risch points out that "there are fundamental differences between the rare, Mendelian disorders and the common 'complex' familial disorders, both in terms of conceptualizing and approaches to analysis, that need to be addressed before significant progress can be made in understanding the 'complex' diseases." Ibid. at 4. See also Eliot Marshall, "Highs and Lows on the Research Roller Coaster," 264 *Science* 1693–1965 (1994); Rose, "The Rise of Neurogenetic Determinism."

132. See Natalie Angier, "Maybe It's Not a Gene Behind Those Wild and Wacky Times," *New York Times*, November 1, 1996.

133. Lawrence K. Altman, "Falsified Data Found in Gene Studies," *New York Times*, October 30, 1996, at A12.

134. Kenneth F. Schaffner, "Complexity and Research Strategies in Behavioral and Psychiatric Genetics," at 13 (paper prepared for Symposium on Biology and Culture, Galveston, Texas, November 1996).

135. Keay Davidson, "DNA: High Hopes or Just Hype?" *Houston Chronicle*, September 29, 1996, at 2.

136. Sandra Blakeslee, "Some Biologists Ask, 'Are Genes Everything?'" *New York Times*, September 2, 1997, at B7, B13.

137. Leon Eisenberg, "The Social Construction of the Human Brain," 152 *Am. J. Psychiatry* 1563, 1569 (November 1995).

138. Ibid. at 1569.

139. Ibid. at 1571.

140. Abby Lippman, "The Genetic Construction of Prenatal Testing: Choice, Consent, or Conformity for Women?" in Karen H. Rothenberg and Elizabeth

Thomson, eds., *Women and Prenatal Testing: Facing the Challenges of Genetic Technology* 9–34, 28 (Columbus: Ohio University Press, 1994).

141. Rothman, *Genetic Maps and Human Imaginations* at 170.

142. Rose, "The Rise of Neurogenetic Determinism" at 381.

143. Ibid. at 382.

144. The study compared 14,427 adopted Danish men with their biological fathers and found no direct inherited tendency, but a correlation between biological fathers and sons in the commission of property crimes.

145. Daniel Goleman, "New Storm Brews on Whether Crime Has Roots in Genes," *New York Times*, September 15, 1992, at C1.

146. Lori Andrews, "Genetic Privacy: From the Laboratory to the Legislature," 1 *Genome Research* 1 (October 1995).

147. Arthur Caplan (presentation at "Genetic Testing for Breast Cancer Susceptibility: The Science, the Ethics, the Future" symposium, San Francisco, November 22, 1996).

8. Which Conceptual Model Best Fits Genetics?

1.   See Benjamin S. Wilfond and Kathleen Nolan, "National Policy Development for the Clinical Application of Genetic Diagnostic Technologies," 270 *J.A.M.A.* 2948–2952 (1993).

2.   *Munro v. Regents of the University of California*, 263 Cal. Rptr. 878 (Ct. App. 1989).

3.   Idaho Code § 5–334; Ind. Code Ann. § 34–12–1–1; Minn. Stat. Ann. § 145.424(1)-(2); Mo. Ann. Stat. § 188.130; N.C. Gen. Stat. § 14–45.1(e); N.D. Cent. Code § 32–03–43; 42 Pa. Cons. Stat. Ann. § 8305; S.D. Codified Laws Ann. § 21–55–1; Utah Code Ann. § 78–11–24.

4.   Idaho Code § 5–334; Minn. Stat. Ann. § 145.424(1)-(2); Mo. Ann. Stat. § 188.130; N.C. Gen. Stat. § 14–45.1(e); 42 Pa. Cons. Stat. Ann. § 8305(a); S.D. Codified Laws Ann. § 21–55–2; Utah Code Ann. § 78–11–24.

5.   See, e.g., Patricia A. Baird and Charles R. Scriver, "Genetics and Public Health," in John M. Last and Robert B. Wallace, eds., *Public Health and Preventive Medicine* 983, 13th ed. (Norwalk, Conn.: Appleton and Lange, 1992).

6.   George Cunningham, "Maternal Serum Alpha-Fetoprotein Screening in California," in J. Fullarton, ed., *Proceedings of the Committee on Assessing Genetic Risks* (Washington, D.C.: National Academy Press, 1994).

7.   Benjamin S. Wilfond and Norman Fost, "The Cystic Fibrosis Gene: Medical and Social Implications for Heterozygote Detection," 263 *J.A.M.A.* 2777–2783, 2781 (1990) (average annual cost $7,500, lifetime costs at least $200,000). Peter T. Rowley, Starlene Loader, Jeffrey C. Levenkron, Charles E. Phelps, "Cystic

Fibrosis Carrier Screening: Knowledge and Attitudes of Prenatal Care Providers," 9 *Am. J. Prev. Med.* 261–263, 261 (1993) (average annual cost for a patient with cystic fibrosis is $10,000; total direct costs for cystic fibrosis patients overall may be $300 million).

8. B. Meredith Burke, "Genetic Testing for Children and Adolescents," 273 *J.A.M.A.* 1089 (1995) (letter).

9. 106 Cal. App. 3d 811, 829, 165 Cal. Rptr. 477 (2d Dist. 1980).

10. Margery Shaw, "Conditional Prospective Rights of the Fetus," 5 *J. Legal Med.* 63 (1984).

11. Recall, however, that even though many women suffer from breast cancer, only 5–10 percent of women with breast cancer have a hereditary form of the disease. NIH News, "Scientists Report New Lead in the Genetics of Breast Cancer" (press release, September 28, 1995).

12. Caryn Lerman, Karen Gold, Janet Andrain, Ting Hsiang Lin, Neal R. Boyd, C. Tracy Orleans, Ben Wilfond, Greg Louben, and Neil Caporoso, "Incorporating Biomarkers of Exposure and Genetic Susceptibility Into Smoking Cessation Treatment: Effects on Smoking-Related Cognitions, Emotions, and Behavioral Change," 16 *Health Psychol.* 87–99, 96 (1997).

13. Ibid.

14. Ibid. at 94.

15. Ibid.

16. Ellen Wright Clayton, "Screening and the Treatment of Newborns," 29 *Hous. L. Rev.* 85, 104 n. 84 (1992).

17. Duchenne muscular dystrophy is a "progressive deterioration of muscles beginning in infancy and leading to death in second or third decade. Inheritance is X-linked recessive." American Academy of Pediatrics, Committee on Genetics, "Newborn Screening Fact Sheet," 83 *Pediatrics* 449–464 (1989).

18. In the state of New York alone, 25,000 women per year are screened for fetal genetic abnormalities. Kimberly Nobles, "Birthright or Life Sentence: Controlling the Threat of Genetic Testing," 65 *S. Cal. L. Rev.* 2081, 2086 (1992). Michael Malinowski notes: "One reason for our acceptance of extensive prenatal genetic screening is that it is being introduced to us through the health profession rather than through a social movement." Michael Malinowski, "Coming Into Being: Law, Ethics, and the Practice of Prenatal Genetic Screening," 45 *Hastings L.J.* 1435, 1453 (1994).

19. See chapter 4.

20. For a description of the technology, see S. Elias, J. Price, M. Dockter, S. Wachtel, A. Tharapel, J. L. Simpson, and K. W. Klinger, "First Trimester Diagnosis of Trisomy 21 in Fetal Cells from Maternal Blood," 340 *Lancet* 1033 (1992); see also Jane Chuen and Mitchell S. Golbus, "Prenatal Diagnosis Using Fetal Cells from the Maternal Circulation," 159 *West. J. Med.* 308 (1993); Richard

Saltus, "Noninvasive Way Is Cited to Deter Down Syndrome in Fetuses," *Boston Globe*, November 12, 1992, at 8.

21.　Down syndrome is caused by extra genetic material on chromosome 21, which results in various malformations and mental retardation. See American Academy of Pediatrics, "Health Supervision of Children with Down Syndrome," 93 *Pediatrics* 855–859 (1994).

22.　Cystic fibrosis is "the most common potentially fatal genetic disease" among Caucasians. "It is caused by a disorder of exocrine glands. Individuals with cystic fibrosis have a variety of physical abnormalities, most serious among them is chronic obstructive lung disease." Office of Technology Assessment, U.S. Congress, *Healthy Children: Investing in the Future*, appendix M, 263 (1988).

23.　Tay-Sachs disease is a fatal neurodegenerative disorder caused by a genetic mutation. It is common among Ashkenazi Jews. See E. C. Landel, I. H. Ellis, A. H. Fensom, P. M. Green, and M. Bobrow, "Frequency of Tay-Sachs Disease Splice and Insertion Mutations in the UK Ashkenazi Jewish Population," 28 *J. Med. Genet.* 177–180 (1991).

24.　For genetic disorders that are treatable after birth, such as phenylketonuria, there would be no real advantage to screening during pregnancy as opposed to after birth.

25.　L. A. Herzenberg, Diana Bianchi, J. Schroder, H. M. Cann, and G. M. Iverson, "Fetal Cells in the Blood of Pregnant Women: Detection and Enrichment by Fluorescence-Activated Cell Sorting," 76 *Proc. Natl Acad. Sci. U.S.A.* 1453, 1455 (1979).

26.　Ibid.

27.　See *Tarasoff v. Regents of University of California*, 131 Cal. Rptr. 14, 551 P.2d 334 (1976).

28.　See *Simonson v. Swenson*, 104 Neb. 224, 177 N.W. 831 (1920).

29.　In the infectious disease cases, if the third party would have gotten the disease anyway, there was no duty to warn. See *Skillings v. Allen*, 148 Minn. 88, 180 N.W. 916 (1921); *Britton v. Soltes*, 563 N.E.2d 910 (Ill. App. 1990).

30.　The outcome of the legal analysis would change very little even if treatment were available. If the disorder at issue could be treated after birth, then testing the newborn infant would be a less restrictive alternative with respect to the woman than prenatal testing. If the disorder needed to be treated while the fetus was in utero, the case for prenatal testing would be stronger but would still fail, since the treatment would likely be more intrusive than the blood test and would invade the woman's bodily integrity and interfere with her right to privacy. Since the woman would be able to refuse the treatment under *In re A.C.*, 573 A.2d 1235 (D.C. Cir. 1987), and *In re Baby Boy Doe*, 632 N.E.2d 326 (Ill. App. Dist. 1994), the state could not show that the testing would accomplish the end of assuring that the fetus was treated.

31. In U.S. Supreme Court cases, the goal of protecting the public treasury has not been found to be superior to that of protecting individual rights. A person's right to travel is recognized as more important than the drain on the welfare system of the state to which he or she moves. See, e.g., *Edwards v. California*, 314 U.S. 160 (1941).

32. For example, in *People v. Dominguez*, 64 Cal. Rptr. 290, 256 Cal. App. 2d 623 (2d Dist. 1967), a pregnant, unmarried woman with two children was convicted of second-degree robbery. As a condition of her probation, the trial judge directed that she not become pregnant without being married, so that the state taxpayers would be spared the burden of caring for illegitimate children. This state interest in saving money was not seen as overriding a woman's interest in childbearing—thus, it is unlikely that such an interest would be considered valid in the genetic testing situation. Moreover, it is unclear that the state could prove, in a cost-benefit analysis, that screening actually did save a sufficient amount of money to justify infringing upon individual choice. Although aborting a fetus with cystic fibrosis, for example, may save society the costs of rearing that child, the overall costs of screening and providing necessary counseling and other services for all pregnant women might be so great that they would exceed the costs of rearing the few affected children whose births the state is trying to prevent. With respect to cystic fibrosis, "it has been estimated that if a national [carrier] screening program were introduced, it would cost $2.2 million for each case of cystic fibrosis avoided." Benjamin R. Sachs and Bruce Korf, "The Human Genome Project: Implications for the Practicing Obstetrician," 81 *Ob. & Gyn.* 458, 459 (1993).

33. See discussion in Lori Andrews, "Prenatal Screening and the Culture of Motherhood," 47 *Hastings L.J.* 967, 999 (1996), regarding parents' liberty interest in childrearing decisions affecting the traits of the children.

34. Lori B. Andrews, Jane E. Fullarton, Neil A. Holtzman, and Arno G. Motulsky, eds., *Assessing Genetic Risks: Implications for Health and Social Policy*, 262–263 (Washington, D.C.: National Academy Press, 1994) (discussing reasons why newborn screening should not be undertaken for cystic fibrosis).

35. Royal College of Physicians' Working Party, *Report on Prenatal Diagnosis and Genetic Screening: Community and Service Implications* (London: RCP, 1989).

36. Angus Clarke, "Is Non-Directive Genetic Counselling Possible?" 338 *Lancet* 998, 999 (October 19, 1991).

37. N. Press and C. Browner, "Collective Fictions: Similarities in the Reasons for Accepting MSAFP Screening Among Women of Diverse Ethnic and Social Class Backgrounds," 8 *Fetal Diagnosis and Therapy* 97–106 (1993).

38. See Kenneth L. Karst, "The Freedom of Intimate Association," 89 *Yale L.J.* 624 (1980).

39. Some marriages have been stopped when the prospective spouse learned of the other partner's genotype.

40. American Society of Human Genetics, "ASHG Report: Statement on Informed Consent for Genetic Research," 59 *Am. J. Hum. Genet.* 471–474 (1996).

41. Andrews et al., *Assessing Genetic Risks* at 276.

42. "NIH Workshop Statement," 8 *Fetal Diagnosis and Therapy* 6–9, 7 (supp. 1, 1993) (emphasis added).

43. Gail Geller, Barbara A. Bernhardt, Kathy Helzlsouer, Neil A. Holtzman, Michael Stefanek, and Patti M. Wilcox, "Informed Consent and BRCA1 Testing," 11 *Nature Genet.* 364 (1995).

44. B. D. Colen, "Proceedings of the Workshop on Inherited Breast Cancer in Jewish Women: Ethical, Legal. and Social Implications," 10 *CenterViews* 7–10, 9 (Spring 1996).

45. Geller et al., "Informed Consent and BRCA 1 Testing" at 364.

46. See chapter 3.

47. See also Jeffrey R. Botkin, "Fetal Privacy and Confidentiality," *Hastings Center Report* 32–39 (September–October 1995), which argues that "notions of fetal privacy and confidentiality can help to define limits to what parents may reasonably learn about their future child."

48. Lori Andrews, "Genetics and Informed Consent," 271 *Science* 1346–1347 (March 8, 1996) (letter).

49. Ellen Wright Clayton, Karen K. Steinberg, Muin J. Khoury, Elizabeth Thomson, Lori Andrews, Mary Jo Ellis Kahn, Loretta M. Kopelman, and Joan Weiss, "Informed Consent for Genetic Research on Stored Tissue Samples," 274 *J.A.M.A.* 1786–1792 (1995).

50. R. F. Weir and J. R. Horton, "DNA Banking and Informed Consent: Part 1," 17, no. 4 *IRB* 1–2 (July–August 1995); R. F. Weir and J. R. Horton, "DNA Banking and Informed Consent: Part 2," 17, no. 5 *IRB* 1–8 (September–December 1995).

51. Peter S. Harper, "Appendix: Research Samples from Families with Genetic Diseases: A Proposed Code of Conduct," in Angus Clarke, ed., *Genetic Counselling: Practice and Principles* 86–94, 87 (London and New York: Routledge, 1994).

52. Jeffrey P. Struewing, Dvorah Abeliovich, Tamar Peretz, Naaman Avishai, Michael M. Kayback, Francis S. Collins, and Lawrence C. Brody, "The Carrier Frequency of the BRCA1 185delAG Mutation Is Approximately 1 Percent in Ashkenazi Jewish Individuals," 11 *Nature Genet.* 198–200 (October 1995).

53. *Moore v. Regents of the University of California*, 51 Cal. 3d 120, 271 Cal. Rptr. 146, 793 P.2d 479 (July 1990).

54. See, e.g., Alan L. Otten, "Researchers' Use of Blood, Bodily Tissues Raises Questions About Sharing Profits," *Wall Street Journal*, January 29, 1986.

55. Lori Andrews and Dorothy Nelkin, "Whose Body Is It Anyway? Disputes Over Body Tissue in a Biotechnology Age," 350 *Lancet* 53 (1998).

56. Ibid.

57. *Moore v. Regents of the University of California,* 51 Cal. 3d 120, 271 Cal. Rptr. 146, 793 P.2d 479 (1990).

58. This should include extensive Fourth Amendment protections and assurances of probable cause before samples are taken.

59. Harold J. Krent, "Of Diaries and DNA Banks: Use Regulations Under the Fourth Amendment," 74 *Tex. L. Rev.* 49–100 (1992).

60. Jean E. McEwen and Philip R. Reilly, "Stored Guthrie Cards as DNA 'Banks,'" 55 *Am. J. Hum. Genet.* 196, 200 (1994).

61. Del. Code Ann. tit. 16, § 1221(a); Fla. Stat. ch. 760.40; Nev. Rev. Stat. § 629.151; N.M. Stat. Ann. § 24–21–4; N.Y. Civ. Rights § 79–1; Vt. Stat. Ann. tit. 18, § 9332.

62. Del. Code Ann. tit. 16, § 1221(b)(5); Nev. Rev. Stat. § 629.151(4); N.M. Stat. Ann. § 24–21–4(c)(9); Vt. Stat. Ann. tit. 18, § 9332(5)(d).

63. For example, Asch notes that Applebaum and Firestein's *A Genetic Counseling Casebook* "failed to include information about life with the disabling condition under consideration." Adrienne Asch, "Reproductive Technology and Disability," in Sherrill Cohen and Nadine Taub, *Reproductive Laws for the 1990s,* 69–124, 89 (Clifton, N.J.: Plenum, 1989).

64. See chapter 6.

65. Ibid.

66. G. Geller, B. A. Bernhardt, T. Doksum, K. J. Helzlsouer, P. Wilcox, and N. A. Holtzman, "Decision-Making About Breast Cancer Susceptibility Testing: How Similar Are the Attitudes of Physicians, Nurse Practitioners, and At-Risk Women," 16 *J. Clin. Oncol.* 2868–2876, 2872 (1998).

67. E. R. Clayton, V. L. Hannig, J. P. Pfotenhauer, R. A. Parker, P. W. Campbell III, J. A. Phillips III, "Teaching About Cystic Fibrosis Carrier Screening by Using Written and Video Information," 57 *Am. J. Hum. Genet.* 171–181 (1995).

68. See, e.g., *Pratt v. Williams,* 658 So.2d 4 (La. 1995); *Solomon v. Hall,* 767 S.W.2d 158 (Tenn. App. 1988).

69. Caryn Lerman and Robert Croyle, "Psychological Issues in Genetic Testing for Breast Cancer Susceptibility," 154 *Arch. Intern. Med.* 609 (March 28, 1994), citing R. Croyle and C. Lerman, "Psychological Impact of Genetic Testing" in R. Croyle, ed., *Psychosocial Effects of Screening for Disease Prevention and Detection* (New York: Oxford University Press, 1995).

70. Geller et al., "Informed Consent and BRCA1 Testing" at 364.

71. Testimony of Caryn E. Lerman, Ph.D., Associate Professor, Georgetown University Medical Center, Washington, D.C., Senate Cancer Coalition, September 29, 1995.

72. One woman undergoing breast cancer testing, for example, said that she chose to be tested because "the unknown is always more frightening than the known." Quoted in Ruth Sorelle, "Gene Screening: Fear of the Find," *Houston Chronicle,* May 13, 1996, at 7.

73. "A woman with no BRCA1 mutations still faces the same base-line 12 percent lifetime risk of breast cancer that any woman faces, and she could be at high additional risk because of inherited mutations in BRCA2 and other loci." Francis S. Collins, "BRCA1—Lots of Mutations, Lots of Dilemmas," 334 *N. Eng. J. Med.* 186–188 (1996).

74. American College of Medical Genetics Storage of Genetic Materials Committee, "ACMG Statement: Statement on Storage and Use of Genetic Materials," 57 *Am. J. Hum. Genet.* 1499–1500, 1499 (1995).

75. Asch, "Reproductive Technology and Disability" at 90.

76. See chapter 7 for discussion of statutes dealing with insurance and employment discrimination.

77. See, e.g., Council on Ethical and Judicial Affairs, "Use of Genetic Testing by Employers," 266 *J.A.M.A.* 1827 (1991).

78. See chapter 6.

79. *Report of the Working Group of the Stanford Program in Genomics, Ethics, and Society on Genetic Testing for Breast Cancer Susceptibility* at 14 (November 11, 1996).

80. Ibid. at 5.

81. Ibid.

82. Ibid. at 14.

83. Ibid. at 10.

84. Ibid. at 9.

85. Andrews et al., *Assessing Genetic Risks* at 136.

86. This is done with Tay-Sachs proficiency testing. Ibid. at 122.

87. Neil A. Holtzman and Michael S. Watson, eds., *Promoting Safe and Effective Genetic Testing in the United States: Final Report of the Task Force on Genetic Testing* 53 (ELSI, September 1997), http://www.nhgri.nih.gov/ELSI/TFGT Final/.

88. Peter Harper, "Appendix: Research Samples from Families with Genetic Diseases" at 90.

89. *Union Pac. Ry. Co. v. Botsford*, 141 U.S. 250, 251 (1891); *Fosmire v. Nicoleau*, 75 N.Y.2d 218, 551 N.E.2d 77, 80–81 (1990).

90. *Matter of Conroy*, 98 N.J. 321, 486 A.2d 1209 (1985).

91. Equal protection concerns might be raised as well. Such testing might be considered to be discrimination based on pregnancy or perhaps even sex discrimination. *Geduldig v. Aiello*, 417 U.S. 484 (1974).

92. *Doe v. Borough of Barrington*, 729 F.Supp. 376 (D.N.J. 1990). "Society's moral judgment about the high-risk activities associated with the disease, including sexual relations and drug use, makes the information of the most personal kind. Also, the privacy interest in one's exposure to the AIDS virus is even greater than one's privacy interest in ordinary medical records because of the stigma

that attaches with the disease." Ibid. at 384. As has been noted in this book, genetic information raises similar risk of stigma and discrimination.

93. See Lori B. Andrews, *Medical Genetics: A Legal Frontier* 190–194 (Chicago: American Bar Foundation, 1987) (discussing common law actions for breach of medical privacy, including actions based on the tort of privacy, breach of contract, malpractice, and breach of fiduciary duty).

94. *Whalen v. Roe*, 429 U.S. 589, 599–600 (1977); *Doe v. City of New York*, 15 F.3d 264 (2d Cir. 1994).

95. It has long been recognized that a blood sample contains more medical information than does a traditional medical record. As Fred Bergmann of the National Institutes of Health pointed out, "The genetic counselor takes a history and puts it in the computer bank. He also takes a blood sample and puts it in the deep freeze. And from the point of view of confidentiality, I would suggest that there is much more information in the deep freeze than in the computer bank, and I think that point should be appreciated by the lawyers and everyone else." Statement of F. Bergmann in B. Hilton, D. Callahan. M. Harris, P. Condliffe, and B. Berkley, eds., *Ethical Issues in Human Genetics: Genetic Counseling and the Use of Genetic Knowledge* 411 (New York: Plenum, 1973).

96. Michael L. Closen, "Mandatory Disclosure of HIV Blood Test Results to the Individual Tested: A Matter of Personal Choice Neglected," 22 *Loy. U. Chi. L.J.* 445, 454–457 (1991); Martha A. Field, "Pregnancy and AIDS," 52 *Md. L. Rev.* 402, 409–413 (1993).

97. Lori B. Andrews, "Informed Consent Statutes and the Decisionmaking Process," 5 *J. Legal Med.* 163, 215–216 (1984). See, e.g., *Cobbs v. Grant*, 8 Cal. 3d 229, 502 P.2d 1, 12, 104 Cal. Rptr. 505 (1976).

98. *Schmerber v. California*, 384 U.S. 757, 771–772 (1966). The blood test in that case was permissible as a "minor intrusion."

99. *Barlow v. Ground*, 943 F.2d 1132 (9th Cir. 1991), *cert. denied* 505 U.S. 1206 (1992).

100. *Glover v. Eastern Nebraska Community Office of Retardation*, 867 F.2d 461 (8th Cir.), *cert. denied* 493 U.S. 932 (1989).

101. *Union Pac. Ry. Co. v. Botsford*, 141 U.S. 250, 251 (1891); *Fosmire v. Nicoleau*, 75 N.Y.2d 218, 551 N.E.2d 77, 80–81 (1990).

102. *Matter of Quinlan*, 70 N.J. 10, 355 A.2d 647 (1976), *cert. denied* 429 U.S. 922 (1976). *Superintendent of Belchertown State School v. Saikewicz*, 373 Mass. 728, 370 N.E.2d 417, 426 (1977). *Cruzan v. Director, Mo. Dep't of Health*, 497 U.S. 261, 278 (1990). ("A constitutionally protected liberty interest in refusing unwanted medical treatment may be inferred from our prior decisions.")

103. *Roe v. Wade*, 410 U.S. 113 (1973); *Planned Parenthood of Southeastern Pennsylvania v. Casey*, 505 U.S. 833, 859 (1992); *Lifchez v. Hartigan*, 735 F.Supp. 1361 (N.D. Ill. 1990), *aff'd without opinion, sub nom., Scholberg v. Lifchez*, 914 F.2d 260 (7th Cir. 1990), *cert. denied* 498 U.S. 1069 (1991).

104. *In re Baby Doe,* 632 N.E.2d 326 (Ill. App. 1st Dist. 1994). *In re A.C.,* 573 A.2d 1235, 1252 (D.C. App. 1990). See also Rodney Halstead, "A Pregnant Woman's Right to Refuse Medical Treatment—Is It Always Her Choice?" 24 *Creighton L. Rev.* 1589 (1991).

105. Andrews, "Prenatal Screening and the Culture of Motherhood" at 973–988.

106. See, e.g., *Schmerber v. California,* 384 U.S. 757, 771–772 (1966).

107. *In re A.C.,* 573 A.2d 1235, 1246 n.10 (D.C. App. 1990). See also *In re Baby Boy Doe,* 632 N.E.2d 326, 333 (Ill. App. 1st Dist. 1994).

108. See 45 C.F.R. § 46.110 (1991).

109. Office for Protection from Research Risks, National Institutes of Health, *Protecting Human Research Subjects: Institutional Review Board Guidebook* 5–45 (1993).

110. *Moore v. Regents of the University of California,* 51 Cal. 3d 120, 271 Cal. Rptr. 146, 793 P.2d 479 (1990).

111. See, e.g., *Moore v. Regents of the University of California,* 51 Cal. 3d 120, 271 Cal. Rptr. 146, 793 P.2d 479 (1990).

112. Ibid.

113. See, e.g., *Browning v. Norton Children's Hospital,* 504 S.W.2d 713 (Ky. 1974).

114. See, e.g., *Curran v. Bosze,* 566 N.E.2d 1319 (1990).

115. See, e.g., *Brotherton v. Cleveland,* 923 F.2d 477 (6th Cir. 1991).

116. W. French Anderson (presentation at Conference on Mammalian Cloning, Washington, D.C., June 1997).

117. Clarke, "Is Non-Directive Genetic Counselling Possible?" at 998.

118. Bernadine Healy, "BRCA Genes—Bookmaking, Fortunetelling, and Medical Care," 336 *N. Eng. J. Med.* 1448–1449, 1449 (1997).

119. Clarke, "Is Non-Directive Genetic Counselling Possible?" at 999.

120. Ibid.

121. T. C. Brown, D. C. Wertz, R. C. Fox, and B. Lerner, "Ethical Perspectives of Genetic Counselors: Does Area of Specialty Matter?" (abstract, National Society of Genetic Counselors 1999 Education Conference).

122. Sherman Elias and George J. Annas, "Generic Causes for Genetic Screening," 330 *N. Eng. J. Med.* 1611 (June 1994). They also oppose screening for a breast cancer or colon cancer gene in the context of reproduction.

123. Clarke, "Is Non-Directive Genetic Counselling Possible?" at 998.

124. Ibid.

125. Press Release, *Harper's* Magazine, November 12, 1997.

126. AP, "Clinton Seeks More Money to Reduce Gap in Wages," *New York Times,* January 31, 1999, at 19.

127. Daniel J. Kevles and Leroy Hood, "Reflections," in Daniel J. Kevles and Leroy Hood, eds., *The Code of Codes: Scientific and Social Issues in the Human Genome Project* 300–328, 326 (Cambridge: Harvard University Press, 1992).

128. Under the existing fundamental rights analysis with respect to abortion, the U.S. Supreme Court has taken the position that federal constitutional law does not require the public funding of abortion. *Maher v. Roe*, 432 U.S. 464 (1977). Some state legislatures, however, have enacted laws funding abortion for poor women, and some courts have held that the state constitution requires funding of abortion to enable women to exercise their fundamental right to privacy to make reproductive decisions. See, e.g., *Moe v. Secretary of Admin.* 417 N.E.2d 387 (Mass. 1981).

129. Karen H. Rothenberg and Elizabeth J. Thomson, "Women and Prenatal Testing: An Introduction to the Issues," in Karen H. Rothenberg and Elizabeth J. Thomson, eds., *Women and Prenatal Testing: Facing the Challenges of Genetic Technology* 1–4, 3 (Columbus: Ohio State University Press, 1994).

130. Andrews, "Prenatal Screening and the Culture of Motherhood," 973–988.

Index